W9-AXH-402

Bird Flight Performance

Frontispiece. *Immature magnificent frigatebird. Both wing span and wing area are higher in frigatebirds, relative to body mass, than in any other seabird. Frigatebirds soar in thermals over tropical seas, spending weeks or months aloft, as they cannot alight on the water. In addition to their well-known piratical habits, they snatch flying fish and squid as they jump from the sea surface. Puerto Rico.*

Bird Flight Performance

A Practical Calculation Manual

C. J. Pennycuick

Maytag Professor of Ornithology
University of Miami

with photographs by the author

OXFORD NEW YORK TOKYO
OXFORD UNIVERSITY PRESS
1989

Oxford University Press, Walton Street, Oxford OX2 6DP
Oxford New York Toronto
Delhi Bombay Calcutta Madras Karachi
Petaling Jaya Singapore Hong Kong Tokyo
Nairobi Dar es Salaam Cape Town
Melbourne Auckland
and associated companies in
Berlin Ibadan

Oxford is a trade mark of Oxford University Press

Published in the United States
by Oxford University Press, New York

British Library Cataloguing in Publication Data
Pennycuick, C. J.
Bird flight performance
1. Birds. Flight.
I. Title
598.2 1852
ISBN 0–19–857721–4

Library of Congress Cataloging in Publication Data
Pennycuick, C. J. (Colin J.)
Bird flight performance : a practical calculation manual / C.J. Pennycuick, with photographs by the author.
p. cm. Includes index. Bibliography: p.
1. Birds—Flight—Mathematics—Handbooks, manuals, etc. 2. Birds-Flight—Data processing—Handbooks, manuals, etc. I. Title.
QL698.7.P46 1989 598.2′1852—dc19 88–38249
ISBN 0–19–857721–4

Set by
Oxford Text System
Printed by St Edmundsbury Press
Bury St Edmunds, Suffolk

For Sandy and Adam

Preface

This book has its roots in the many hours I have spent watching birds in various parts of the world, from the ground, from powered aircraft, and from gliders. Every birdwatcher wonders how fast and high birds fly, how far they can go, and how they set about it. Of course, the most straightforward way to answer those questions is to observe birds, and see what they do—but what are the problems really, from the point of view of the bird? Pilots view birds with their own special brand of anthropomorphism, since birds and pilots so obviously have to cope with many of the same problems. I have tried to apply elementary aeronautics to birds' problems, and at the same time to avoid letting the discussion degenerate into an arcane form of witchcraft, accessible only to high priests with supercomputers. The non-mathematical reader may flick through the following pages, and think this book is not for him, but please look again. The algebra is there for those who want it, but anyone who can run a BASIC program on a microcomputer can do these calculations. With a pre-recorded program disc, there is not even the pain of typing in the programs.

I have been stimulated to put this book together by many signs of interest among ornithologists who want to know how much fuel birds are likely to consume when migrating or foraging, and so on. The programs will answer such numerical questions, but the computer cannot insist that you believe the answers. It is easy to make performance predictions with the programs, but not so easy to devise field or laboratory observations, that can be used to check the accuracy of the results. I hope that ingenious users will think up diverse ways of checking the output of the programs against observations, as it is only thus that faith in the reliability of the calculations can be established in the long term. Meanwhile, any group of students can collect their own measurements of birds' wings and bodies, and feed them into the programs on a rainy afternoon, to find out which birds are good at different aspects of flight performance. This may be a limited form of experimentation, but it leads to endless surprises. Familiar birds are seen with new eyes.

My thanks are due to the many colleagues, students, fellow pilots, and other friends with whom I have watched birds and argued about flight over the years. I should not single anyone out, except that I must record my special thanks to Mark Fuller for the endless hours he has spent going through earlier versions of this material, and trying to teach me how to present it in an intelligible form. If I have failed, it is not his fault, but I certainly could not have got as far as I did without his help. I hope that someone, somewhere will find this book and its programs useful, or at least have some fun with them.

Miami 1988 C.J.P.

Contents

A note on the disc

The program disc supplied with this book contains the three BASIC programs in Appendix 1, and an 'Infofile' with brief instructions for running the programs. Chapter 4 contains further information about the programs, and advice on transferring them to computers running operating systems other than MS-DOS. Users who have problems running the programs should contact the author at: Department of Biology, University of Miami, PO Box 249118, Coral Gables, Florida 33124, USA.

Key to bird silhouettes

p. 1; Magnificent frigatebird, Panama. p. 7; Black-browed albatross, South Georgia. p. 18; Arctic skua, Shetland. p. 30; Brown pelican, Florida. p. 41; Gannet, Shetland. p. 64; Andean condor, Peru. p. 81; Southern skua, South Georgia. p. 95; Turkey vulture, Peru.

1. The mechanical approach to flight energetics

Introduction

Many biologists believe that the most direct way to estimate the rate at which an animal uses energy in locomotion is to measure its oxygen consumption, or to measure fuel consumption by some physiological method. Actually millilitres of oxygen, and grams of fat, are not units of energy. To convert such observations into estimates of energy, assumptions have to be made about the energy released when the substrate is oxidized. Also, physiological measurements relate to the total of all processes in the animal that consume energy, and it is difficult to divide this total up, in a way that allows the actual work done by the muscles to be estimated. Physiological measurements are at their best in elucidating the mechanisms by which energy is mobilized and used for locomotion. Their most serious drawback is their inherent lack of generality. A measurement of, say, the oxygen consumed by a pigeon flying in a wind tunnel, applies to that particular bird under the conditions of the experiment, but cannot be transferred to predict the energy consumption of other species, flying under other conditions. Some authors have tried to find a basis for such predictions, by assembling measurements of 'flight metabolism' from the literature, and performing multivariate regressions against various body measurements. Sadly, such endeavours are doomed to failure, for reasons that will become clear.

The aeronautical approach

A vast literature exists on the principles of flight, much of it unfortunately expressed in a mathematical idiom which most biologists find obscure. The older literature deals with the physical problems that had to be overcome in the development of low-speed aircraft—essentially the same ones that were also overcome in the evolution of flying animals. For several decades, beginning in the early years of the twentieth century, these problems were the field of interest of the world's aeronautical research organizations, and their accumulated publications today occupy sizable sections of libraries. Around the time of World War II, the centre of interest shifted to speeds and altitudes far above those inhabited by animals, but the earlier works on 'incompressible' aerodynamics remain as valid and useful today, as they were when they were written. The principles were encapsulated in several famous textbooks, of which the most revered are those of Prandtl and Tietjens (1934*a*, *b*) and Milne-Thomson (1958). Other favourites are von Mises (1945) and Abbott and Doenhoff (1959). These classics are still in print, thanks to that indispensible publishing house Dover Publications, and are widely used by engineers and students around the world. An important addition to the main body of classical theory was supplied by Schmitz (1960), who extended it to cover the sizes and speeds of model aircraft, which are also those of birds.

The engineers' approach is exactly opposite to that of the physiologists. They cannot begin by measuring an aircraft's fuel consumption, because there is no aircraft to study until their labours are complete. They begin instead by studying the physical properties of air, and the way that it behaves (on both microscopic and macroscopic scales) when flowing around objects of various shapes. Then, they study the forces produced on different shapes under various conditions of flow. Then, they design complete structures that should generate the forces they want, under the conditions they have in mind. They test structures or parts of structures, in flight or in wind tunnels, to find out whether the forces agree with those predicted, and if not, they look for ways to improve the basis of prediction. Fuel consumption only comes when the design process is nearly enough complete to supply a thrust force, which can be multiplied by speed to get the power required to propel the finished aircraft. The most important characteristic of this method of analysing flight performance is that it is general, in that it can be applied to aircraft of any size or shape. Its application to flying animals has recently been reviewed by Norberg (1989).

Natural selection in engineering

While the basic principles are simple enough, the actual calculations often get quite intricate. Biologists often think that these calculations are highly 'theoretical', but actually this is not so. Aeronautical theory has been shaped by a form of natural selection, just as rigorous as that which orchestrates organic evolution. When the designer's work is done, his plane has to fly, and it has to perform more or less as predicted. In engineering, wrong theory does not survive.

Origin of this book

The purpose of this book is to explain how classical aeronautical theory can be adapted for predicting the flight performance of birds, and also to supply the reader with a practical toolkit, consisting of BASIC programs that can be entered and used by anyone with access to a microcomputer. These programs are introduced in Chapters 4 and 6. The theory embodied in Programs 1 and 1A (for level flapping flight) originated in an early adaptation of aeronautical theory by Pennycuick (1969). This version of the theory had a number of shortcomings, many of which were pointed out by Tucker (1973), who published an expanded version of the theory. However, there are some difficulties with Tucker's (1973) version, pointed out in a further revision by Pennycuick (1975). This 1975 version, which incorporated most of Tucker's (1973) amendments, was presented in a form that was intended to enable readers to make their own performance calculations, with no more computing power than a slide rule. A large class of practical problems, concerned with energy requirements for migration or foraging flights, can be addressed with this theory, but it has been disappointingly little used in the years since its publication. More regrettably, some predictions, allegedly derived from it, are difficult to reconcile with the data supplied. I hope that the computer programs will overcome any inhibitions that readers may have about making their own calculations. If the user's copy of a program reproduces the output from the test examples correctly, then it should run correctly on other data.

Power required and power available

Power required to fly

Most of this book is concerned with calculating how much power is needed to keep a bird flying level at some steady speed (V). Chapter 3 explains in detail how this is done. If the calculated power is plotted for various values of the speed, a curve with a well-marked minimum results, as shown in Fig. 1.1. There is a definite value of the speed, known as the *minimum power speed* (V_{mp}), at which less power is required to fly, than at either faster or slower speeds. Also marked on Fig. 1.1 is a higher speed, labelled V_{mr}. This is the *maximum range speed*. It is the speed at which the ratio of power to speed is least, and can be found by drawing a tangent to the power curve as shown. A bird flying at this speed requires more power than one flying at V_{mp}, but has to do less work per unit distance flown, than at any other speed. A bird should fly at V_{mp} if it wants to minimize the amount of work done (or fuel consumed) per unit *time*. If the requirement is to minimize the work done per unit *distance* (or to fly as far as possible before the fuel is exhausted), then V_{mr} is the optimum speed at which to fly.

The curve in Fig. 1.1 is only drawn from a little below V_{mp} to a little above V_{mr}. This is deliberate. The assumptions underlying the theory given

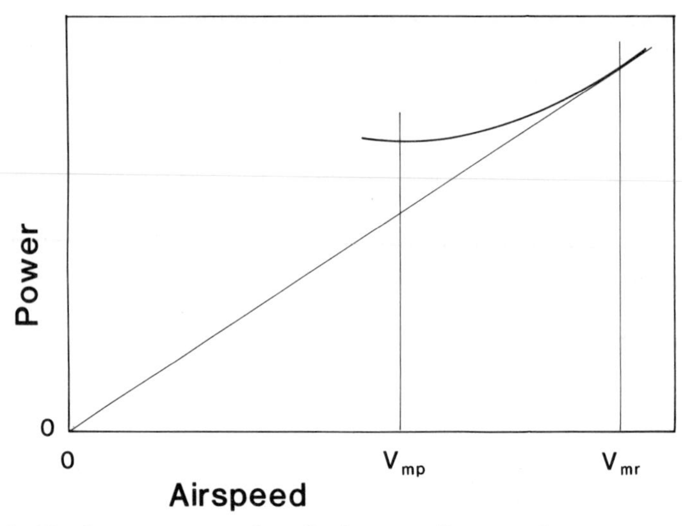

Fig. 1.1. *Total power required to fly horizontally, as a function of speed. The minimum power speed* (V_{mp}) *and maximum range speed* (V_{mr}) *are explained in the text.*

here become unrealistic for very low or very high speeds, i.e. much below V_{mp} or above V_{mr}. The computer programs described in Chapter 4 automatically select lower and upper limits of speed, beyond which this method of calculation should not be extrapolated. Chapter 5 gives a qualitative explanation of the reasons why other considerations have to be taken into account outside this speed range, particularly in very slow flight and hovering. This book will not serve as a 'cookbook' for problems involving very low-speed flight, but other sources are indicated in Chapter 5.

Power available from the muscles

If a bird is observed to be capable of flying level, it follows that its flight muscles are able to produce at least the amount of power needed to fly level at V_{mp}. Most birds can produce more power than this, since they can fly either faster or slower than V_{mp}, take off, climb, and perform other manoeuvres that call for more power than is needed to fly at V_{mp}. Chapter 7 outlines the principles that determine how much power is available from the muscles, and gives enough information to enable the reader to make numerical estimates. No computer programs are supplied, but the calculations are so simple that a pocket calculator is sufficient. The reasoning is based entirely on mechanical considerations. The reader who believes that muscle power output is determined by such factors as enzyme activities will find another point of view in Chapter 7. The biochemical systems have to be adapted to supply energy at a rate at which the contractile machinery can use it. The upper limit to this rate, on a mass-specific basis, is lower in large animals than in small ones, and is determined by mechanical factors only.

Gliding and soaring

The power requirements for gliding flight can be calculated in a very similar way to those for flapping flight. The difference is that in gliding, the power comes from the bird using up its potential energy (that is, sinking earthwards), rather than from its muscles. In soaring, the loss of potential energy is offset by flying in rising air, or extracting energy from air movements in some other way. Soaring is the only type of flight where a pilot can fly alongside a bird, and be reasonably sure that both are using the same source of energy in the same way. The sport of gliding is much concerned with the speed at which a glider can make progress across country (and not at all with the amount of noise it makes). Birds also soar to cover distance on foraging flights, and on migration. Chapter 6 explains the basis of soaring performance, and Program 2 does the calculations for a bird of the reader's choice.

Uses and limitations of the programs

Programs 1, 1A, and 2 represent the bird in terms of a small number of body measurements, notably the mass, wing span, and wing area. If you supply the measurements, the programs will do the calculations without enquiring whether the numbers came from a bird, a bat, an insect, an aircraft, or a pterosaur. The empirical data, such as they are, that have been incorporated as default values in the programs, come mainly from medium-sized and large birds. Caution is needed in extrapolating too far down the size range, to small birds, bats, and insects, for reasons that are indicated in qualitative terms in Chapter 5. The flight of birds, bats, and pterosaurs has been reviewed in detail by Norberg (1989), and there do not appear to be major differences that would invalidate the calculations, provided the limitations of the programs are observed. The most important of these, as noted above, is that the two powered-flight programs (1 and 1A) are intended for steady, level flight at cruising speeds between the minimum power speed and the maximum range speed—not for hovering, slow flight, manoeuvring, or maximum-speed flight as in raptors in pursuit of prey. Similarly, Program 2 applies to gliding with the wings set for minimum drag, in the range of speeds used for inter-thermal glides, as discussed in Chapter 6. Provided these limitations are understood and borne in mind, the reader can enter the measurements of his eagle, duck, guillemot, or albatross, and expect to get sensible estimates of speed, power, and fuel consumption. Confidence in the absolute values of the predictions will improve as users devise new ways to check the predictions against field or laboratory measurements, and adjust the default values of variables as necessary. Some examples of ways in which this can be done are given in Chapter 8. The programs have been designed to facilitate 'what-if?' calculations, in which the program is run several times, changing the values of one or more variables between each run. The percentage changes of speed and power that result can usually be believed with more confidence than the absolute values. A class of students can try a wide range of variations, using their own or other measurements.

Inevitably the output of the programs will have to be treated with increasing scepticism the further the user's scenario is removed from the empirical base of bird flight measurements. So long as that is understood, there is no need to feel inhibited about calculating flight performance, merely because the circumstances have never been observed. For example, calculations on the flight performance of penguins are quite enlightening even though hardly realistic (Pennycuick 1987). Again, an example in Chapter 8 uses Program 2 to predict the cross-country performance of a warbler soaring in thermals. The results do not have to be believed to a high degree of precision to see why warblers do not use this method for their migrations. The reader may find it amusing to fly familiar birds on planets whose atmospheres and gravities differ from those here on Earth, or to invent flying animals that might reasonably be expected to exist, but have escaped the notice of taxonomists. The programs will produce estimates of speed and power, but will not indicate whether the flight muscles would have any hope of generating the power required, or of delivering it to the wings without exceeding the strengths of bone and tendon.

2. Variables needed for flight calculations

Introduction

Mechanical calculations depend on measuring physical quantities of various kinds, and combining them to derive other kinds of quantities. For example, work (energy) is calculated by measuring forces and distances, and multiplying them together. All biologists have learned at some stage in their lives that work equals force times distance, and most can recite this fact correctly. However, when it comes to measuring energy, many biologists treat it as though it were an entirely separate type of quantity, to be expressed in whatever arbitrary units happen to come to hand, and do not seem to realize that it has any connection with force and distance. The kinds of variables needed for mechanical calculations are very simple, but it is essential that the physical character of every variable (its 'dimensions'), and the relationships between different variables, are explicitly recognized and continually kept in mind.

Dimensions and units

The reader who is unfamiliar with the system for representing dimensions in terms of mass, length, and time (M, L, and T), will find it worth studying. These formulae are a shorthand that defines the physical nature of each quantity, that is whether it is a length, an area, a pressure, an energy, or whatever. One of their more straightforward uses, illustrated in the following chapters, is for checking that calculations make sense in physical terms. For example, if you multiply a length by an area, the result has to be a volume, and if it purports to be, say, an energy, then something must be wrong. The expressions on opposite sides of the 'equals' sign in an equation must both have the same dimensions, and where several terms are added together, the dimensions of each term must be the same.

Identifying the dimensions of a quantity is a matter of physical accuracy, whereas selecting units in which to measure it is a matter of convention. For example, the metre, the foot, the furlong, and the parsec are all units of length, and if lengths are the only type of quantity that you wish to measure, then it is a matter of convenience which you choose. However, if you wish to combine measurements of length with other measurements involving mass and time, then it helps to have a system or family of units, that are related to each other in a simple manner. For example American engineers measure distances in feet, forces in pounds, and masses in a unit called the slug, which few non-engineers have even heard of. This system may look eccentric, but it is internally consistent, in the sense that if you multiply a mass (in slugs) by an acceleration (in ft s^{-2}), the resulting force comes out in pounds. The SI system, which all scientists are supposed to use, also has this property of internal consistency. The calculations in this book are unlike most in biology, in that they combine physical quantities of several different kinds (dimensions), to create others that have different dimensions again. Such calculations would be a nightmare without an internally consistent system of units.

The calorie is an obsolete unit of heat, which does not belong to any system of units.

Standard Measurements

Certain measurements have to be made on the bird, and on the air, in order to run the computer programs in this book. To run Programs 1 and 1A (for powered flight), the minimum information you need is the bird's mass and wing span. You need the wing area in addition to run Program 2 (for gliding). The programs will supply values for all other variables, although you will probably wish to change some of these 'default values' to try out hypotheses, or to use your own measurements. The most reliable way to get body measurements for some particular bird is to measure it yourself. There are some collections of data in the literature (e.g. Greenewalt 1962), but they need to be treated with caution, because ornithologists have proved surprisingly creative in inventing wrong ways to make even the simplest measurements. Worse, they often do not specify exactly how the measurement was made. Some of the more minor deviants are illustrated in Figs. 2.2 and 2.3. The reader can see that several of these could make a large difference to the result. Some also involve an element of judgement, so that repeated measurements, even by the same observer, cannot be relied upon to give the same result. The methods of measuring wing span and area given below have been designed primarily to make the measurements repeatable between different observers. If you use these methods to make your own measurements, then please say so when publishing the data—if not, please give similarly precise details of your own methods.

Mass (m) *and weight* (mg)—*dimensions M and MLT*$^{-2}$

The familiar operation of 'weighing' a bird is a prolific source of confusion and misunderstanding. When you wrap a bird in a bag and suspend it from a spring balance, the instrument registers its weight, that is, the force that gravity exerts on it, dragging it towards the centre of the earth. Most ornithologists express the result in grams, and think of it as a measure of the amount of material making up the bird's body. However, this 'quantity of material' concept is the mass, not the weight. A spring balance can be used to measure mass, but only if you know the acceleration due to gravity. If we assume standard earth gravity of 9.81 m s^{-2}, then the weight of a bird whose mass is 1 kg would be

$$1\ \text{kg} \times 9.81\ \text{m s}^{-2} = 9.81\ \text{kg m s}^{-2}.$$

These units, kg m s^{-2}, may look strange, but they are the correct units of force (and therefore of weight) in the SI system. The unit also has a shorter name, the 'newton', abbreviated 'N'. The conversion factor between the mass and the weight of a bird is not necessarily 9.81, but depends on the strength of gravity. For instance, if you could take your 1 kg bird to the moon, its mass would still be 1 kg, but it would weigh only 1.62 N, because gravity is weaker on the moon. The spring balance would indicate the reduced weight correctly, but if you wanted it to read mass directly in grams, you would have to recalibrate it for moon gravity. The best way to avoid confusion is to represent the weight as mass multiplied by gravity. If the mass is represented by m, and the acceleration due to gravity is g, then the weight is mg. The quantity that you measure on the bird is always the mass, and you record it in grams or kilograms. On those occasions when you actually want the weight, you multiply the mass by g, to get the weight in newtons.

The newton is not just a smaller unit than the kilogram. It is a different type of unit, with different physical dimensions. The dimensions of all quantities used in this book can be expressed as combinations of three fundamental quantities. In the SI system, these quantities are chosen (arbitrarily to some extent) to be mass, length, and time, abbreviated M, L, and T. The dimensions of mass are simply M. Force is obtained by multiplying mass (dimensions M), by acceleration (dimensions LT^{-2}). Therefore the dimensions of force (including weight) are MLT^{-2}, and the units as above. A summary of dimensions and units is in Appendix 2. More extensive information, with tables of conversion factors for a wide selection of physical quantities, may be found in Pennycuick (1988).

Wing span (b)—*dimensions L*

Wing *span* is the most important morphological measurement required on the bird—actually it is the only one, apart from the mass, that is absolutely necessary for calculations on powered flight. It is quick and easy to measure. If field ornithologists would just remember to record and tabulate wing spans in their publications, a useful body of data could quickly be built up. Wing span is defined as the distance from one wing tip to the other, with the wings spread horizontally as far out as they will go (Fig. 2.1). This measurement has to be made on a living or freshly dead bird. It cannot be made on a museum skin or skeleton. If the measurement has to be made on a bird with one damaged wing, a second-best method is to measure from the centreline of the back (not the shoulder joint) to the tip of the good wing, and double the result. Once again, the wing must be stretched out as far as it will go. Right and wrong ways of measuring wing span are illustrated in Fig. 2.2.

Wing area (S)—*dimensions* L^2

Wing *area* is needed in calculations about gliding flight. It is more troublesome to measure than wing span, but field ornithologists should try to collect measurements, especially on soaring birds, if they have the opportunity. Wing area is defined as the projected area of both wings, fully spread out, including the area of that part of the body that is included between the wing roots (Fig. 2.1). The simplest way to measure it is to

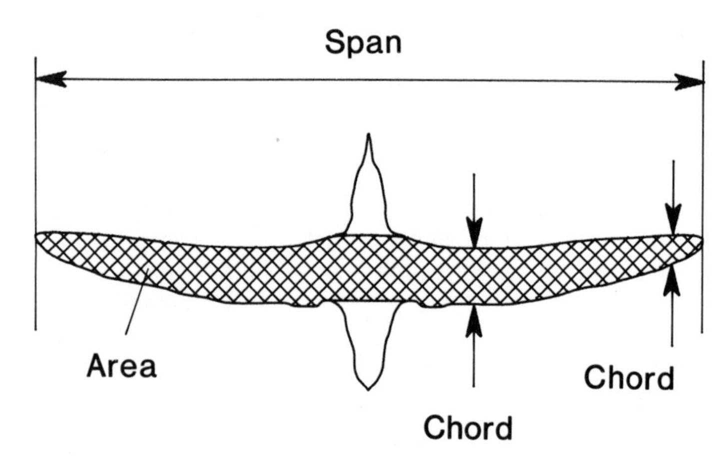

Fig. 2.1. *Wing span and area are defined as shown, with the wings spread out to the sides to their fullest extent. Wing chord is the width of the wing, measured along the direction of flight. It varies at different points along the span.*

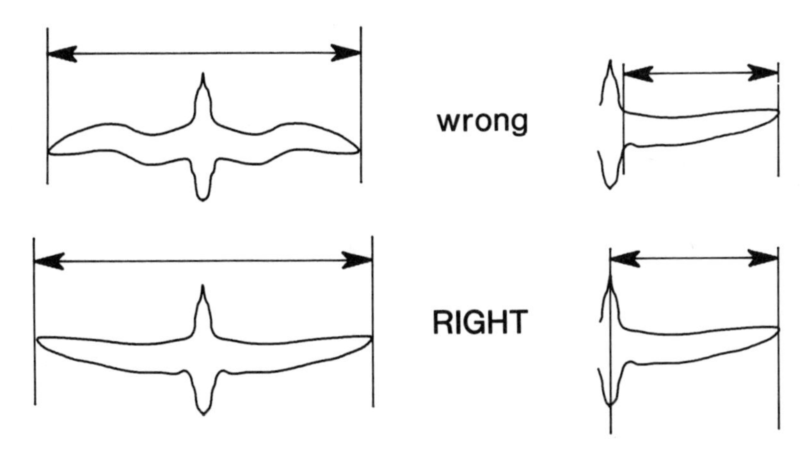

Fig. 2.2. *Right and wrong methods of measuring wing span.*

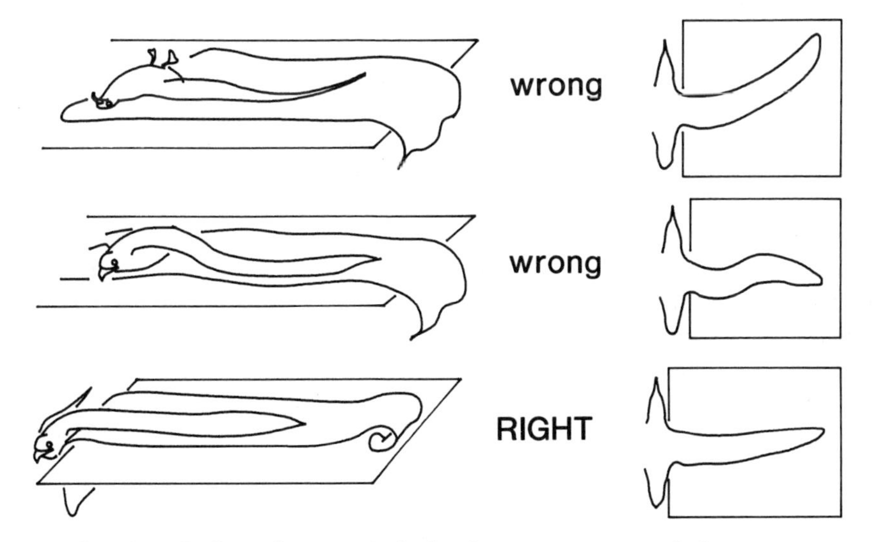

Fig. 2.3. *Right and wrong methods of tracing a wing to find its area.*

make a tracing of one wing as shown in Fig. 2.3. A sketchbook can be used for tracing small birds, while larger ones can be traced on a roll of brown parcel paper, or on sheets of newspaper taped together as necessary. The bird is held right-way-up with the wing spread over the paper on a board or table. The wing should be spread straight out sideways to its fullest extent, and the body should be held *beside* the edge of the table (not on it), so that the wing rests flat on the drawing surface. Trace around the outside of the wing, following the outline of the individual feathers. It is not necessary to define the proximal end of the wing.

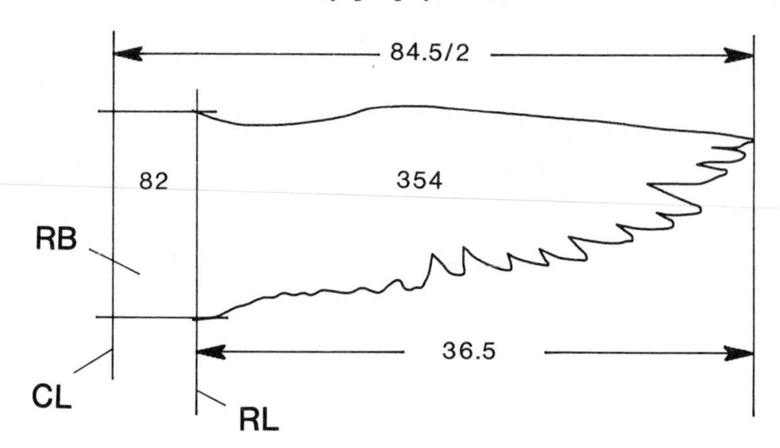

Fig. 2.4. *Finding wing area from the tracing (see calculation in text). CL, centreline; RL, root line; RB, root box.*

To find the area from the tracing, first rule off the proximal end with a straight line as shown in Fig. 2.4. This 'root line' should be parallel to the body centreline, but its exact location is not important. Then measure the area enclosed by the wing, distal to the root line. The best way to do this is to overlay the tracing with a grid ruled on a transparent plastic sheet, and count the squares. 1 × 1 cm squares are precise enough for all but very small birds. Around the edges, count those squares that are more than half filled, and exclude those that are not.

The result of this measurement is an incomplete area of one wing. The method of completing it is shown in Fig. 2.4. You have to extend the proximal end of the wing inwards to the centreline of the body, adding a 'root box', that extends from the centreline to the root line. Adding the area of the root box to the area of the enclosed outline gives the total area for one side. Doubling this gives the total wing area. In the example of Fig. 2.4 (an oystercatcher, found dead), the wing span was recorded as 84.5 cm, and the area of the wing distal to the arbitrary root line was found by gridding to be 354 cm^2. The calculation of wing area then proceeds as follows:

This wing:	354 cm^2
Root box:	82 cm^2
Total this side:	436 cm^2
Both sides:	872 cm^2
	= 0.0872 m^2

To find the area of the root box, you have to measure inwards from the tip to the root line (36.5 cm) and subtract this distance from half the span. Do not be misled by the additional decimal places that appear on the calculator when you do this. Whatever the calculator says, you cannot

Fig. 2.5. *An immature red-tailed hawk showing wing-tip slots, formed by emarginated primaries. Some ornithologists think that it is tidier to measure the wing area after joining the tips of the primaries by straight lines. However, the bird's weight is supported by the feathers, not by the gaps between them, so measure the area of the feathers, and omit that of the gaps. Puerto Rico.*

measure the wing area of an oystercatcher-sized bird to the nearest square millimetre, so round the numbers. Three significant digits are enough. A calculator that rounds numbers automatically is a great convenience. Notice that if the root line had been drawn a little further out (perhaps because of a damaged feather), the area listed as 'This wing' would have been smaller, but the root box would have been bigger by approximately the same amount. Thus variations in the position chosen for the root line make little or no difference to the result.

Wing chord (c)*—dimensions L*

The wing *chord* is the width of the wing, from the leading edge to the trailing edge, measured along the direction of flight (Fig. 2.1). In a tapered wing, the chord is large near the wing root, and diminishes towards the tip. The chord at any particular point on the wing also changes as the bird flexes or extends its elbow and wrist joints. The *mean chord* of the fully spread wing, used for calculating Reynolds numbers, is found by dividing the wing area by the span. Some ornithologists use the term 'chord' to refer to the distance from wrist to wing tip on the folded wing. Such misapplication of technical terms causes a great deal of unnecessary confusion. The wrist-to-tip measurement is not the chord, and should be given a different name, such as 'wrist-to-tip distance' or 'folded wing length'. It does not have any straightforward significance for flight mechanics.

Aspect ratio (Λ)*—dimensionless*

The *aspect ratio* is a simple measure of the shape of a wing. It is the ratio of the wing span to the mean chord. Since both of these are lengths, the aspect ratio is dimensionless. It is usually calculated as

$$\Lambda = b^2/S. \tag{2.1}$$

As with span and area, the aspect ratio should be recorded with the wing at full stretch. Gliding birds reduce their aspect ratios in fast flight by flexing the wrist and elbow joints.

Wing loading (W)*—dimensions* $ML^{-1}T^{-2}$

Wing *loading* is another quantity that is useful for calculations about gliding. It is the weight (not mass) per unit wing area, in other words the mean pressure difference between the lower and upper surface of the wing when the bird is in a steady glide. The dimensions are the same as those of pressure. Wing loading is calculated as:

$$W = mg/S. \tag{2.2}$$

The SI units are newtons per square metre. This unit is also called the pascal (Pa).

Body frontal area (S_b)*—dimensions* L^2

Body *frontal area* is the cross-sectional area at the widest part. The computer programs in this book automatically estimate it from the body mass, but also allow the user to substitute a measured value. The easiest way to measure frontal area is to find the body circumference, by wrapping a dressmaker's tape measure around the body, tightly enough to compress the feathers to about the degree expected in flight, at the widest point behind the wings. The cross-sectional shape of most bird bodies is not very different from circular, and the frontal area can be estimated on that basis from the circumference (C) as:

$$S_b = C^2/4\pi. \tag{2.3}$$

Air density (ρ)*—dimensions* ML^{-3}

The *air density* obviously has a major effect on the forces acting on a wing, and the work required to flap it. The symbol ρ is a Greek 'rho' (not 'p', which stands for pressure). The ambient air density should always be observed and recorded in conjunction with field or laboratory measurements on bird flight (a precaution seemingly unknown to flight physiologists). Observers on the ground usually measure the barometric pressure and the temperature, then find the air density (in kg m^{-3}) from the formula:

$$\rho = 1.23 \times (p/p_0) \times (288/(T + 273)) \quad (2.4)$$

where T is the air temperature in degrees Celsius, p is the ambient barometric pressure, and p_0 is the barometric pressure at sea level in the standard atmosphere. p_0 is 101 300 pascals, or 1013 millibars, or 29.92 inches of mercury. Use whichever value matches the units on the barometer. For airborne observations, pressure is usually measured with the aircraft's altimeter, which is actually an aneroid barometer, although it is calibrated to show altitude, rather than indicating pressure directly. The indicated altitude (h) should be recorded with the altimeter set to 1013 mb (or 29.92 in. Hg), and the outside air temperature should also be noted. If h is expressed in feet, the ambient pressure can then be found from the formula:

$$p = p_0 (1 - (6.88 \times 10^{-6})h)^{4.256} \quad (2.5)$$

This value for the pressure can then be used in equation (2.4) above. Of course, the same method can be used on the ground, if you happen to have an aircraft altimeter.

Values for the air density at various altitudes in the standard atmosphere are in Appendix 4, but it should be remembered that actual atmospheric conditions at any particular height can deviate quite widely from the values chosen for the standard atmosphere. It is best to use observed values to go with field observations. This is easy enough to do so long as the observer is at the same height as the bird. High-precision meteorological instruments are not necessary. A cheap aneroid or electronic barometer and a regular thermometer are quite sufficient. The computer programs will work out the air density from the readings of these instruments. Further instructions for entering the air density into the computer programs are in Chapter 4.

Kinematic viscosity (ν)*—dimensions* L^2T^{-1}

Kinematic viscosity is defined as the ratio of the viscosity of a fluid to its density. Its dimensions turn out to be L^2T^{-1}, so the SI units are m^2s^{-1}. Values for air at various heights in the standard atmosphere are in Appendix 4. They are needed for calculating Reynolds number (Chapter 5).

Work, power, and metabolic rate

Work (Q)*—dimensions* ML^2T^{-2}

Program 1 (described in Chapter 4) works out how much fat a given bird will consume in flying a given distance. To do this, it first estimates how much energy is required to fly horizontally for a given distance. Energy is work, that is force times distance. In effect, the program works out the average forward force (thrust) that the bird has to exert to keep itself moving forwards at a steady speed, and multiplies this by the distance flown. The product of the force times the distance is the amount of work that has to be done by the flight muscles. In the SI system, force is measured in newtons (above), and distance in metres, so the work comes out in newton metres, also known as joules (J). Its dimensions are ML^2T^{-2}.

Measurements of oxygen consumption provide an indirect means of estimating the amount of energy that an animal consumes. The energy estimated by such methods is identical physically with ordinary mechanical work. Its dimensions are ML^2T^{-2}, the same as those of work, because it is work. Heat, and energy released by chemical reactions, result from forces at the molecular level, moving their points of application through some distance, in exactly the same way as the macroscopic forces that propel a large animal. James Joule demonstrated in 1843 that heat and work are physically the same, and it follows that if you already have units for force and distance, you do not need a separate unit for work. The calorie, an arbitrary and archaic unit of heat, was rendered obsolete by Joule's discovery, although this fact remains to this day largely unnoticed in the world of physiology. To judge from the labels on their graphs, some physiologists seem to believe that the millilitre of oxygen is a unit of energy.

When an animal proceeds in steady aerobic locomotion, the performance of a certain amount of work by its muscles, as measured by the force they exert and the distance through which they shorten, leads eventually to the consumption of a corresponding amount of fuel, and of oxygen. Estimating just how much fuel and oxygen correspond to each joule of mechanical work involves a number of assumptions, which come in two stages. First, the performance of one joule of mechanical work by a muscle always requires the consumption of more than one joule of fuel energy. The ratio of work done to fuel energy consumed is the *conversion efficiency* (η). This is dimensionless, being a ratio of work:work. Second, various biochemical assumptions have to be made, in order to estimate what mass of fuel must be consumed, to liberate the required amount of fuel energy. This is expressed as the *energy density* of the fuel. Its dimensions are those of energy divided by mass, i.e. L^2T^{-2}.

In Program 1, the default value assumed for the conversion efficiency is 0.23, that is, out of each joule of fuel energy consumed, 0.23 J is assumed to be converted into mechanical work by the muscles, while the remaining 0.77 J appears as heat. It is further assumed that aerobic bird flight is fuelled by fat, and that the energy density of fat is 3.9×10^7 J/kg. These values can be amended by the user when the program is run, and the default values can easily be changed by doctoring the program.

Power (P)—*dimensions* ML^2T^{-3}

Power is the rate of doing work. As the SI units of work and time are the joule and the second, the unit of power is the $J\ s^{-1}$, also known as the watt (W). Physiological experiments measure the rate of consumption of fuel or oxygen, from which an indirect estimate of power can be obtained if assumptions are made about the substrate, and the amount of energy released when it is oxidized. For some reason physiologists like to describe estimates of power, made in this way, as 'metabolic rate'. There is never a valid reason to express such estimates in any units other than watts.

Since work is force times distance, power can also be represented as force times speed. Power (rather than energy) is the quantity calculated first in flight performance calculations. This is done directly in a piecemeal fashion, by estimating the magnitudes of various forces acting on the animal, and multiplying each force by the speed with which it moves its point of application. The total power is built up by calculating various components and adding them together. Finally, fuel consumption can be estimated if assumptions are made about the conversion efficiency and the energy density of the fuel.

3. Power required for horizontal flight

Introduction

A bird proceeding horizontally at a steady speed consumes energy at a steady rate. This chapter presents a practical method for calculating what that rate is, in other words for finding the power, as a function of speed, over a defined range of speeds. The method follows that of Pennycuick (1975), with only minor modifications, which are indicated. BASIC Programs 1 and 1A, described in Chapter 4, calculate the power (and some extras) by the method given here. The differences between the two versions, and their different purposes, are explained later in this chapter. It is possible to use the programs without reading the present chapter, although some understanding of what each program is doing will help to make the output more readily intelligible.

Synthesis of the power curve

The calculation proceeds by identifying several different processes that require the bird to expend power. The power required for each is estimated, as a function of speed. These individual components are then added together to give the total power, which can be plotted on a graph, with forward speed as the *x*-axis.

Parasite power (P_{par})

The easiest component of power to understand and calculate is the *parasite power*, which is the power needed to propel the bird's body (excluding the wings) through the air. The name 'parasite power' was coined by aero engineers who consider power 'parasitic', if it is needed to overcome the drag of parts that do not contribute to supporting the weight. If a bird is flying along horizontally at a steady speed, not accelerating in any direction, it follows from Newton's first law of motion that the resultant of all the forces acting on it must be zero. One of these forces is the aerodynamic

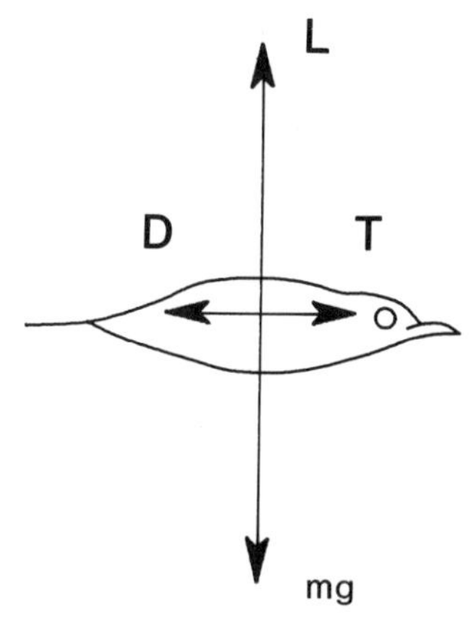

Fig. 3.1. *Balance of forces on the body in level flight. Over a whole number of wingbeat cycles, an average lift force* (L) *balances the weight* (mg), and an average thrust (T) *balances the drag* (D) *of the body and wings.*

drag of the body, a force which acts backwards along the flight path. If this were the only horizontal force, the bird would decelerate. Since it does not, the drag must be balanced by a forward force of equal magnitude (the thrust), which is generated in some roundabout way by the bird flapping its wings. Exactly how the bird achieves this need not concern us. If the bird neither accelerates nor decelerates, it follows that the horizontal components of all the various pushes and shoves administered by the wing to the body, in the course of one complete wingbeat cycle, must average out to a thrust force (T), part of which balances the body drag (D_{par}) as shown in Fig. 3.1. The parasite power is then the product of this component of the thrust and the forward speed:

$$P_{par} = D_{par}V. \tag{3.1}$$

To estimate the parasite power from equation (3.1), you first have to find the parasite drag. The see how this is done, first imagine a tube with an open end pointing into the relative wind, and the other end connected to a manometer. Pressure builds up in the tube, because air molecules (which have mass) approach at a speed V, and decelerate to zero speed as they are stopped by the stationary air in the tube. Each molecule is decelerated by a force, directed upstream, which amounts to saying that there is a pressure gradient, with the pressure in the tube being higher than that far upstream. The pressure difference between the tube and the free stream is the 'dynamic pressure' (p), and its magnitude is:

$$p = \tfrac{1}{2}\rho V^2. \tag{3.2}$$

If it is not self-evident that multiplying a density by a velocity-squared will result in a pressure, we can check the dimensions. ρ has the dimensions ML^{-3}, and V^2 has L^2T^{-2}. Multiplying these together gives $ML^{-1}T^{-2}$, which are indeed the dimensions of pressure (Appendix 2). Note that the dimension check ascertains that both sides of the equation have the same physical character, but it does not check that the magnitudes are also the same. The factor 1/2, being dimensionless, does not affect the dimensions.

Pressure is force per unit area. Therefore the force (drag) exerted on the tube is the pressure (p) multiplied by the cross-sectional area of the tube. The drag (D) on a plate of area A, which brings the wind completely to a stop, is $D = pA$. The value of A, which results in the same amount of drag as that produced by the bird body, is said to be its *equivalent flat-plate area.* The actual frontal area of the body (S_b) is greater than A, because the body is streamlined, and only slows the wind down a little, without bringing it to a stop. If the bird body produced the same amount of drag as a flat plate with 20 per cent the cross-sectional area (say), then the equivalent flat-plate area would be $0.2S_b$. The dimensionless multiplier (0.2 in this case) is known as the *drag coefficient* of the body (C_{Db}). Thus the parasite drag is:

$$D_{par} = \tfrac{1}{2}\rho V^2 A = \tfrac{1}{2}\rho V^2 S_b C_{Db} \tag{3.3}$$

and the parasite power is the drag multiplied by the speed:

$$P_{par} = \tfrac{1}{2}\rho V^3 S_b C_{Db} \tag{3.4}$$

Two numbers have to be known about the bird in order to estimate its parasite drag from equation (3.4); the frontal area of its body (S_b) and its drag coefficient (C_{Db}). The frontal area can be measured as indicated in Chapter 2, or it can be estimated from the formula:

$$S_b = (8.13 \times 10^{-3})m^{0.666} \tag{3.5}$$

where the mass (m) is in kilograms and the frontal area is in square metres. This formula was derived empirically from measurements on various waterfowl and raptors by Pennycuick, Obrecht, and Fuller (1988), and may be used for other birds unless actual measurements are available. Body drag coefficients were measured in the same study, and by Prior (1984). They show a tendency to be less in large, fast birds, than in small, slow ones. The reasons for this are discussed in Chapter 5. It is difficult to make accurate drag coefficient measurements, but 'best guess' estimates are in the region of 0.25 for birds like geese and swans, 0.40 for small birds, and intermediate for pigeon-sized birds. Formulae published by Pennycuick, Obrecht, and Fuller (1988) for calculating both frontal area (equation (3.5) above) and drag coefficient are built into the computer programs. The programs will automatically calculate both quantities from the body mass, but will also give you an opportunity to substitute your own measured values if you have them. This procedure replaces the method for estimating A given by Pennycuick (1975), which does not take account of the variation of C_{Db} with size and speed. This is the only part of the computer programs that differs from the earlier procedure, and the effect of the modification on the results is minor.

Induced power (P_{ind})

Most objects accelerate earthwards when dropped in air, but a bird in horizontal flight does not do this. It follows that its weight, like the drag of the body, must be balanced by an equal and opposite force (Fig. 3.1). It is uncertain who first identified the origin of this large 'lift' force, but the correct solution to the puzzle was known to nineteenth-century theorists such as von Helmholtz and Lord Rayleigh, and was one of the most important elements that turned ancient dreams of human flight into a branch of engineering science. The lift force is created by pushing on molecules of air (which have mass), and making them accelerate downwards. A downward force is applied to the air, and the air responds by accelerating downwards, so that work is done in moving it. The rate at which this work has to be done is the power required to support the weight, and is known as the *induced power* (P_{ind}).

Force, as noted above, is the same as mass multiplied by acceleration, but it can also be seen as the rate of change of momentum. Momentum is mass multiplied by velocity (dimensions MLT^{-1}). The dimensions of the rate of change of momentum are therefore MLT^{-2}, the same as those of force. Rate of change of momentum is in fact the same thing as force. We know the force with which the bird is pushing downwards on the air (it is equal to the weight), therefore we can work out the rate at which momentum is imparted to the air (also equal to the weight). This in turn allows the induced power to be calculated.

Figure 3.2 shows a simplified way of looking at this. The bird is flying along at a speed V, but if we arrange for it to do this in a wind tunnel, the bird can be stationary, with the wind blowing past it at V, as shown, which is more convenient for for observation. The bird flaps its wings in some manner whose details need not concern us, with the result that a part of the airstream that has been influenced by the wings acquires a downward component of velocity, which it did not possess before it passed the bird. We have to determine the rate at which downward momentum is acquired by the air, and then set this equal to the bird's weight. To get the rate of change of momentum, we first find the rate at which mass is flowing past the bird (dimensions MT^{-1}), and then multiply it by the downward velocity imparted (LT^{-1})—but how much mass is influenced by the wings in each second? At first sight the answer seems indeterminate, because air at some

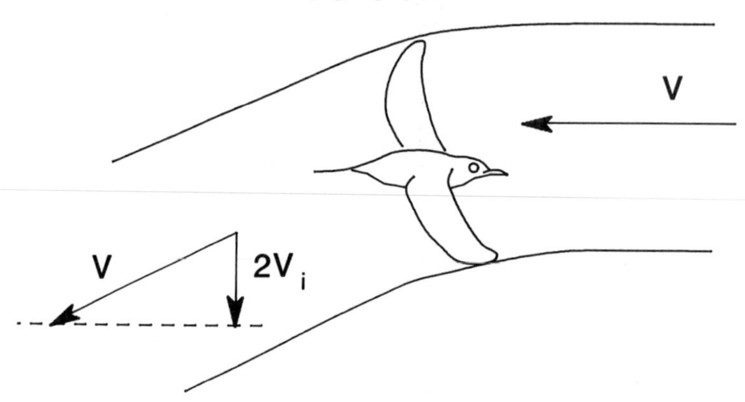

Fig. 3.2. *The 'momentum jet' theory of lift generation. By the time the air reaches the wing disc, it has acquired a downward velocity equal to* V_i *(the induced velocity). It continues to accelerate downward behind the bird, reaching an eventual downward velocity of* $2V_i$.

distance from the flight path might be deflected by just a small amount. The problem can be handled with the help of a surprising simplification, discovered by F. W. Lanchester around the turn of the twentieth century. It can be assumed, as shown in Fig. 3.2, that there is a tube of air, within which all the air is accelerated to the same downward velocity, while the air outside the tube is not affected. The cross-sectional area of this tube is the same as that of a circle, whose diameter is the wing span. It might be thought that only the air in the cross-section actually swept by the wings, as they flap up and down, would be deflected, but this is not so. The angle through which the wings are flapped does not affect the cross-section of the tube of accelerated air. The tube occupies the full circle, *even for a fixed wing*, which does not flap at all. This result may seem improbable, but it has been verified by decades of experimental aeronautical research.

Lanchester's theorem makes it easy to work out the rate of change of momentum. We are concerned with the rate at which air flows through a circle, whose diameter is equal to the wing span. The area of this circle is called the *disc area* (S_d), where

$$S_d = \pi b^2/4. \tag{3.6}$$

Multiplying this area (dimensions L^2) by the speed V at which the air is flowing through it (dimensions LT^{-1}), gives the volume rate of flow (dimensions L^3T^{-1}). This can then be multiplied by the air density (ρ—dimensions ML^{-3}) to get the rate at which mass flows through the disc (dimensions MT^{-1}). If we call the mass flow F_m, then

$$F_m = S_d V \rho. \tag{3.7}$$

To get the rate at which the air acquires momentum, the mass flow has to be multiplied by the downward velocity to which it is ultimately accelerated, far downstream of the bird. We do this in terms of the *induced velocity* (V_i), which is the downward component of velocity that the air has acquired, by the time it passes through the wing disc. However, the air has not finished accelerating at the time it reaches the wing disc. Its acceleration up to that point is caused by suction above, and ahead of the wing disc. After traversing the disc itself, the air enters a region of increased pressure, below and behind the wing, which causes it to continue accelerating, until eventually its downward velocity reaches $2V_i$, far behind the bird. Now we can multiply the mass flow (equation (3.7)) by $2V_i$, and set the result equal to the weight:

$$mg = 2V_i(S_d V\rho). \tag{3.8}$$

All the variables in this equation can be measured, except for V_i, so we can turn it round to find V_i:

$$V_i = mg/2(S_d V\rho). \tag{3.9}$$

We know the force with which the bird is pushing downwards on the air (mg), and we now also know the speed at which it has to push it (V_i). The power, being force times speed, is:

$$P_{ind} = km^2g^2/2(S_d V\rho). \tag{3.10}$$

The factor k in equation (3.10) is a dimensionless *induced drag factor*. It is assigned a default value of 1.2 in the computer programs. It is introduced because the simplified theory given above (for which $k=1$) represents an ideal situation. The real situation deviates from the ideal in ways that are indicated qualitatively in Chapter 5, and any such deviation results in more induced power. There are indications that the default value of $k=1.2$ may be too high for cruising flight (Chapter 8). The user can change the value of k when running the programs.

Parasite and induced powers combined

There are more components of power to be calculated, but already some important characteristics of the power curve can be seen by adding together P_{par} and P_{ind}, that is, by considering how much power the bird needs to support its weight, and to overcome the drag of its body at the same time. If you plot a curve of P_{ind} against V from equation (3.10), you will find that you cannot begin at zero speed (try it), but once your bird is moving forwards, the curve drops away as shown in Fig. 3.3. Some people find it surprising that the cost of supporting the weight declines as speed increases. It does so because there is plenty of air rushing past at high speeds, so that a large force can be produced without having to do much work on the air.

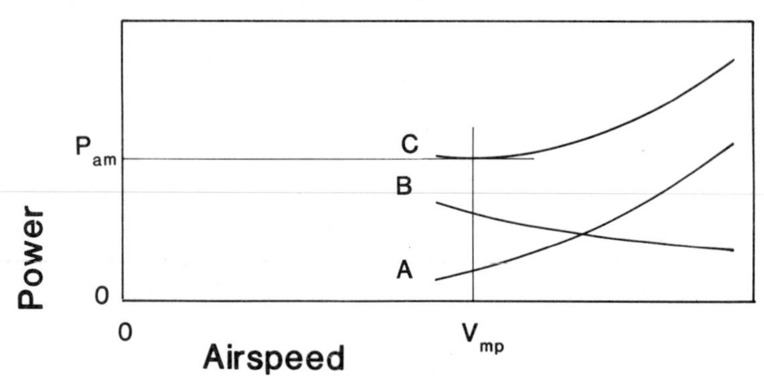

Fig. 3.3. *A, parasite power as a function of speed; B, induced power; C, the sum of parasite and induced powers. Curve C has a minimum value* (P_{am}) *at the minimum power speed* (V_{mp}). *This does not occur where induced and parasite powers are equal, as sometimes stated, but at the point where the* slopes *of curves A and B are equal and opposite.*

The curve of P_{par} starts at zero when the speed is zero, and remains barely perceptible for a while, then curves ever more steeply upwards as the speed increases (Fig. 3.3). If you add P_{ind} and P_{par} together (not starting at zero speed), the curve slopes downwards at first, because P_{ind} makes the biggest contribution to the total. Then, as P_{ind} dwindles away, and P_{par} starts to build up, the curve of total power passes through a minimum, and continues to bend upwards at an ever steeper gradient. The existence of a *minimum power speed* (V_{mp}), at which flight is less strenuous than at either faster or slower speeds, is characteristic of any animal or machine that flies by imparting downward momentum to the air. V_{mp} can be estimated without much error from the curve of parasite plus induced powers, as in Fig. 3.3. The sum of the parasite and induced powers at this speed is called the *absolute minimum power* (P_{am}). Of course the bird cannot actually fly with as little power as P_{am}, because there are further components of power that have not yet been considered. Although P_{am} is a theoretical value, it can easily be calculated (see Pennycuick 1975), and serves as a useful 'yardstick' for assessing the magnitudes of other components of power.

Profile power (P_{pro})

The third mechanical component of power is the *profile power*, which is less easy to understand than the induced or parasite powers, and impossible to calculate with any degree of exactness. We have to resort to simplifications and approximations. Profile power is a portion of the power that has to be supplied by the flight muscles to sweep the wing through its arc from high to low and back again. We have already taken account of some of the work that the muscles do in calculating the induced power (above), that is, we have allowed for that portion of the work which is converted into kinetic energy of air molecules, as they are accelerated downwards to generate the lift force. However, even if the bird sets its wing in such a way that no lift is produced, the wing still produces some drag, and work has to be done in rotating it against this drag. This residual drag (not due to generation of lift) is the profile drag, and the power needed to overcome it is the profile power.

Just as parasite power depends on the speed at which the air blows over the body, so the profile power depends on the speed at which the air blows over the wing, but the latter is not so easy to define. The relative airspeed 'seen' by each portion of the wing is due to two main components, the bird's forward speed, and the relative speed due to flapping. At high speeds, the forward speed predominates, and this affects all parts of the wing equally, but at low speeds, the more distal parts of the wing 'see' a higher relative airspeed than the parts near the body, because they are being rotated through a longer arc. The local component of speed due to flapping, at any particular point on the wing, is itself a function of forward speed, because the amplitude and frequency of flapping are high at very low speeds, then decline to minimum values somewhere near V_{mp}, then build up again as speed is increased further. Profile power has to be calculated in a piecemeal fashion, by dividing the wing into a number of strips, each with its local relative airspeed, taking account of variations of forward speed, and of flapping frequency and amplitude. Then, you need a *profile drag coefficient*, corresponding to the body drag coefficient of equation (3.3). Some authors seem to believe that the profile drag coefficient can be regarded as constant, but unfortunately aeronautical research indicates that it is strongly dependent on the angle at which the relative airflow strikes the wing, especially in strongly cambered wings like those of birds, and especially under the conditions that prevail in flight at very low speeds.

Enough has been said to show that the calculation of profile power is very complicated, and also requires values to be assumed for variables that are inaccessible to any practicable method of measurement. The results of my own most recent attempt at this type of calculation may be found in Pennycuick (1968*b*), and did not supply much encouragement to try this method again. However, this particular effort brought to light an interesting property of the profile power, which can be exploited as a simple method of approximation. In hovering, the wings beat rapidly through a large angle, but the relative airspeed is still quite low, because only the induced velocity supplies the air flowing through the wing disc. The profile power therefore starts low, when the forward speed is zero, but builds up sharply as the speed increases. As the forward speed approaches V_{mp}, the component of profile power due to forward speed builds up, but the component due to flapping declines, because the frequency and amplitude of flapping decline, and so also does the profile drag coefficient. The two effects roughly balance, so that the profile power stays steady, or even declines slightly, as the speed comes up to V_{mp}, and does not increase appreciably until the speed is well above V_{mp}. Eventually, at very high speeds, the curve of profile power would bend upwards in much the same way as that of parasite power, although it is unlikely that many birds have enough muscle power to find that out.

This behaviour of the profile power suggests a very simple way in which it can be represented for the purposes of estimating flight performance. So long as attention is confined to a limited range of speeds, starting just below V_{mp}, profile power can be regarded as constant, that is, independent of the forward speed. We still need a means to determine its magnitude, for some bird that is known to us only in terms of its body mass and its wing span. The 'absolute minimum power' (P_{am}, illustrated in Fig. 3.3) depends on the mass and the wing span, and can easily be calculated. Further, it was argued by Pennycuick (1975), that any changes of anatomy that increase P_{am} should also increase P_{pro} by the same factor; in other words, that P_{pro} can be considered a fixed multiple of P_{am}. The ratio of P_{pro}:P_{am} is called the *profile power ratio* (X_1). It is dimensionless. A default value of 1.2 is used in the computer programs, but once again there are indications that this may be too high (Chapter 8).

This 'quick and dirty' method of estimating P_{pro} is implemented in Programs 1 and 1A. It makes the calculation very simple, but it is essential to remember that its justification only applies over a limited range of speeds. This is one reason why the calculation is not recommended outside the range of speeds that is selected automatically in the programs.

Metabolic power (P_{met})

One of the shortcomings of an earlier version of this procedure (Pennycuick 1969) was that it took account of the above three mechanical components of power only. This is correct for some types of calculation, but if the objective is to estimate how much fuel a bird uses in prolonged aerobic flight, then it is also necessary to take account of basal metabolism. The notion behind this is that basal metabolism is a constant power 'overhead' that is required at all times, whether the bird is flying or at rest. To account for it, we estimate the basal metabolic rate, and simply add it to the other demands for power that have already been calculated. The basis for doing this is the regression equations derived by Lasiewski and Dawson (1967). The power estimated from these equations is of course a rate of consuming *chemical* energy, whereas the components already calculated are in mechanical form (see Chapter 2). To make it compatible with the mechanical components of power, the basal metabolic rate is multiplied by the conversion efficiency to give its 'mechanical equivalent'. This is called the *metabolic power* (P_{met}).

Circulation and respiration power

P_{par}, P_{ind}, P_{pro}, and P_{met} are the four 'direct' components of power in aerobic flight. However, there is a further 'overhead' that also applies only in aerobic flight, the need for which was pointed out by Tucker (1973). In aerobic locomotion, fuel and oxygen have to be transported to the muscles, and carbon dioxide and heat have to be transported away, and disposed of. Tucker proposed that the rate of blood-flow needed to do this is proportional to the power. Pumping this blood calls for further expenditure of power by the heart muscles. The muscles that pump air in and out of the lungs also have to expend power, presumably at a rate that increases with increasing demand for oxygen. The basis for estimating the magnitude of these power requirements is slender indeed. Tucker (1973) proposed simply adding 10 per cent to the total power required for other purposes. Actually it is more likely that this component of power would vary with the square of the total direct power, but so little information is available that Tucker's suggestion may be favoured, on the grounds of its simplicity. It is incorporated in the computer program, in the form of the factor R, whose default value is set to 1.1.

Total power in aerobic and anaerobic flight

We can now put together the total power required for horizontal flight. If the bird's exertions are anaerobic, only the mechanical components of power apply, and the total required is:

$$P_{an} = P_{par} + P_{ind} + P_{pro} \tag{3.11}$$

whereas in aerobic flight this becomes

$$P_{aer} = R(P_{par} + P_{ind} + P_{pro} + P_{met}). \tag{3.12}$$

The fundamental difference between Programs 1 and 1A is that the 'power' calculated by Program 1 is P_{aer} from equation (3.12), whereas Program 1A finds P_{an} from equation (3.11). Program 1 is used for problems that involve fuel or oxygen consumption, while Program 1A is for problems in which mechanical performance is estimated or observed directly, without any complications due to the conversion of fuel energy into work.

Fuel consumption in relation to distance

Power is work done per unit time. When considering the energy requirements of foraging or migrating birds, we are often more interested in the work done (or fuel consumed) per unit distance. This is equal to the power divided by the speed (try the dimensions). Work can be converted into fuel energy required, or mass of fat consumed, as discussed in Chapter 2.

V_{mp} is not the most economical speed at which to fly, in terms of fuel consumed per unit distance. The ratio $V{:}P_{aer}$ reaches a maximum at a higher speed than this, known as the *maximum range speed* (V_{mr}), which may be found by drawing a tangent to the power curve from the origin, as shown in Fig. 3.4. It may seem paradoxical that the bird can cruise more economically by flying faster than V_{mp}, even though it has to work harder to do so. It is true that more power is required to fly at V_{mr} than at V_{mp}, but in changing from one to the other, the speed increases by a bigger factor than the power. Above V_{mr}, this is no longer true. Program 1 calculates and tabulates the fat consumption (per unit distance) over a range of speeds. This reaches a minimum at V_{mr}, whereas the power reaches a minimum at V_{mp}.

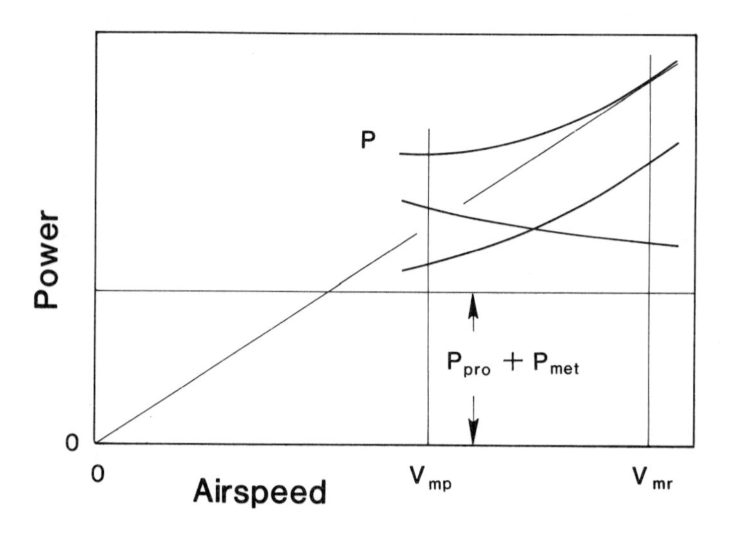

Fig. 3.4. *The same graph as Fig. 3.3, displaced upwards by a fixed increment, equal to the sum of profile and metabolic powers. The value of* V_{mp} *is unaffected by this. The top curve* (P) *is the total power, found by adding together* P_{par}, P_{ind}, P_{pro}, *and* P_{met}. *The maximum range speed* (V_{mr}) *can now be found by drawing a tangent from the origin, as shown.*

Effective lift:drag ratio

At any particular speed, the *effective lift:drag ratio* (N) is defined as

$$N = mgV/P. \tag{3.13}$$

N is listed in the output tables of Programs 1 and 1A, using different definitions of P, according to equations (3.12) and (3.11) respectively. The maximum value of N, achieved when V is equal to V_{mr}, is listed at the bottom of the tables as 'L/D max'. The fat consumption (C_f) at each speed is found in Program 1 from the effective lift:drag ratio:

$$C_f = mg/Ne\eta, \tag{3.14}$$

where e is the energy density of the fuel (default value $3.9 \times 10^7\,\mathrm{J\,kg^{-1}}$ for fat), and η is the conversion efficiency. The subsidiary range program of Program 1 finds the range (Y) of a long-distance migrant from the formula

$$Y = (e\eta N/g)\ \ln[1/(1-F)], \tag{3.15}$$

where F is the 'fat fraction', that is, the proportion of the all-up mass at take-off that consists of consumable fuel. The rationale behind this is explained by Pennycuick (1975).

Effect of wind

Until this point, V has always stood for airspeed. If the bird flies against a headwind, its *groundspeed* (V_g) will be less than its airspeed, and vice versa. In calculations about foraging or migration, we are usually concerned with the amount of fuel needed to fly for a given distance over the ground, rather than through the air. Program 1 allows a head or tailwind to be specified, and uses the ratio V_g/P_{aer} to work out the fat consumption.

4. BASIC programs for flight calculations

Introduction

Program 1, described in this chapter, was designed primarily for those who wish to calculate the fuel requirements of a foraging bird, carrying food back to its nest. It also includes an option for estimating the range of a long-distance migrant. Progam 1A is a version that includes mechanical power only, for studies in which muscle performance, rather than fuel consumption, is the point of interest. Much of the general advice in this chapter also applies to Program 2, which is for gliding flight, and is described in Chapter 6. It is, of course, possible to enter data and get output, without having any idea what the programs are doing, but the reader will find it helpful to read Chapter 3 before making serious use of Programs 1 and 1A, and Chapter 6 before using Program 2. Listings of all three programs are in Appendix 1. Those who cannot get a pre-recorded program disc will find some advice later in this chapter on entering the programs from the listings.

Using a pre-recorded program disc

All of the programs require a printer to be connected to the computer. They will not run without one. They also require the computer to have a BASIC interpreter, preferably originating from Microsoft. Advice on dealing with problems that may arise with other versions of BASIC is given later in this chapter. Having checked that the printer has plenty of paper in it, and is switched on, you first have to invoke BASIC. The manual for the particular computer will tell how to do this. When BASIC is ready it says 'OK'. Now you can insert the program disc, and load one or other of the programs. For example, to load Program 1 on an MS-DOS system type:

```
LOAD "Prog_1.BAS"
```

If this does not work, check the directory of the program disc to see what the file name is. Note the use of the string quotes, and the underline character, used instead of a space in the file name. When the program has been loaded successfully, BASIC says 'OK' again. Type RUN to run the program. The program should print a message on the printer giving details of this book. Then it will ask for the following details about the bird.

Empty mass and payload

All the programs call for the body mass in two instalments, 'Empty mass' and 'Payload'. If in doubt, suspend the bird from a balance, enter the result (in kilograms) as 'Empty mass', and enter zero for 'Payload'. The 'Payload' feature is provided for users who want to make a quick comparison of the power requirements for the same bird, carrying different amounts of food. The empty mass is the bird's body mass, with nothing in its crop. In the case of a forager, the payload is the mass of the crop contents, while for a long-distance migrant, it is the mass of stored fat that is available for use as fuel. The mechanical power requirements are calculated on the basis of the all-up mass, that is the sum of the empty mass, payload, and radio transmitter (if any—see below). As far as these components of power are concerned, it is not necessary to know how the mass is distributed between empty mass and payload. The output of Program 1A, which only calculates these components, depends only on the all-up mass, and the 'Payload' input is just a matter of convenience for entering problems that involve a forager carrying various amounts of load.

The output of Program 1 is affected by the distribution of mass between payload and empty mass, because basal metabolic rate is included as a component of the total power, and this is calculated from the empty mass alone. It (and the total power) will be slightly overestimated if the all-up mass is entered in place of the empty mass. In large birds, there will probably be no discernible effect on the estimates, because the basal metabolism is a very small fraction of the total power requirements. In small birds, basal metabolism is a larger fraction of the total power, and there may be a discernible difference in the power requirements, depending on how the mass is assumed to be distributed. This should not be seen as a serious threat to the accuracy of the results. The estimate of basal metabolism is a rough-and-ready procedure at best, being based on the regression equations of Lasiewski and Dawson (1967).

Wing and body measurements

All the programs call for the wing span (in metres), and Program 2 also requires the wing area (in square metres). The body frontal area and drag coefficient are automatically calculated from the mass, although the main menu provides an option to replace the calculated values with measured values if you have them.

The main menu

Having supplied the body measurements, you will be presented with a menu inviting you to change the values of a list of variables, including the ones you have just entered. Each of the variables shown can be changed, by entering the appropriate number from the menu, and pressing 'Return'. You will be prompted for a new value of the variable. When you have supplied one, the menu will reappear, duly amended. The following items often have to be changed from the menu, and require some comments.

Empty mass and payload If you amend the empty mass or payload in between program runs, you also get recalculated values for the body frontal area and drag coefficient, based on the new all-up mass. If you are using your own estimates for frontal area or drag coefficient, you have to enter them again, each time you change the empty mass or payload.

Air density The air density is set to a default value of 1.23 kg m^{-3}, which is the value for sea level in the standard atmosphere. This will not always be appropriate. The air density will be less than the default value at higher altitudes, and it is also affected by variations of barometric pressure and air temperature. The 'standard' density for various altitudes can be found in Appendix 4. The density can be entered directly, or alternatively the user is offered the option of entering an observed air temperature and pressure instead, as explained in Chapter 2. The computer will then work out and enter the air density. The pressure should be read from a barometer at the flight location and altitude, and the reading used as it stands (*not* corrected to sea level). Do not use the barometric pressure from a local weather service, because this will be corrected to sea level (unless your bird is flying at sea level).

Radio transmitter All the programs include an option for adding a radiotracking transmitter. If this option is selected from the main menu, the program will prompt you for the mass of the transmitter (in kilograms), and then for the equivalent flat-plate area. The notion of the equivalent flat-plate area of a bird's body is introduced in Chapter 3 and further discussed in Chapter 5. The idea is that a transmitter that projects from the surface of the body will add an increment to the flat-plate area, although this increment should be less than the actual frontal area of the transmitter. There is some empirical information about this in Obrecht, Pennycuick, and Fuller (1988), suggesting that for a back-mounted radio with end fairings, fully exposed to the airstream, the increment of flat-plate area is about half the actual frontal area of the radio. In the case of a tail-mounted transmitter, completely covered up by the tail coverts, the increment of flat-plate area may be assumed to be zero. The increment of flat-plate area has to be entered in square metres.

Units and number entry The programs require masses, lengths, and areas to be entered in the main SI units, kg, m, and m^2. This may be a nuisance at first for those who are in the habit of recording their masses in g, wing spans in cm, and areas in cm^2 or mm^2. Inconveniently small numbers are often required, especially for areas. However, most versions of BASIC have a convention that makes these conversions painless. For example, if your bird's empty mass is 82.4 g, there is no need to work out how many zeros you need after the decimal point. Just enter 82.4E−3 (no spaces). The computer reads this as 82.4×10^{-3}. Try it on your computer to make sure it works. To convert centimetres into metres, multiply by 10^{-2}, cm^2 to m^2 by 10^{-4}, and mm^2 to m^2 by 10^{-6}. Those who need further help with units and conversion factors will find it in Pennycuick (1988).

Running the program

Once the values shown in the menu are satisfactory, enter the number for 'Run standard program' and press 'Return'. In response to the prompt 'Version?', you can enter 'male', 'female', or whatever, or you can ignore this by pressing 'Return'. Whatever you enter here will be printed out after the species name in the output table. Do not include any punctuation in your response to 'Version?', as this is liable to confuse BASIC. The program should now go through the rest of its sequence without any further intervention on your part, producing an output table like the examples shown in Appendix 1. First, it produces a block of output listing the values being used for certain variables, whose meanings are explained in Chapter 3. Some of these are from your input, but most are default values that can be altered from the main menu. Next, there will be a short pause (or a lengthy wait, depending on your computer), while the computer works out the power curve, printing out a table of results as it does so. Program 1A prints the mechanical power and the effective lift:drag ratio at each speed. In Program 1 the corresponding figures are the total power (including metabolic components of power), and the effective lift:drag ratio based on the total power. This total power is still expressed in mechanical form, but Program 1 has another column in which the total power is divided by the conversion efficiency (η) to give an estimate of the rate of consumption of fuel energy. This column should be used for comparison with the results of physiological experiments based on measuring oxygen or fuel consumption. The last column produced by Program 1 is the amount of fat used per unit distance flown (g km^{-1}). This refers to distance flown *over the ground*, taking into account any head or tail wind that you may have specified from the main menu. Nothing is printed in this column unless the groundspeed is at least 0.5 m s^{-1}. This last column is the only part of the output that is affected by the wind strength.

Finally, the computer prints out values for V_{mp}, V_{mr}, P_{min}, P_{mr}, the best effective lift:drag ratio, and, in the case of Program 1 only, the fat consumption when flying at V_{mr} (expressed as grams of fat consumed per *air* kilometre flown). These variables are explained in Chapter 3. In Program 1, the estimates for P_{min} and P_{mr} represent the total power, including metabolic components, but expressed in mechanical form. They need to be divided by the conversion efficiency (η) to change them into rates of consumption of fuel energy. The wind does not affect any values in this block of output.

If you want to see the effect of varying any of the values in the first block of output, type 'Y' at the prompt 'Do more?' to return to the main menu. Make whatever changes you want from the menu, and run the program again, entering a note at the prompt 'Version?' to identify the current trial. It is easy to do several runs, systematically changing the values of one or more variables.

Fuel consumed on short and long flights

The numbers in the last column of the power curve table produced by Program 1 can be used to estimate the amount of fat used on short flights (up to 100 km or so), on which the amount of fat consumed is small in relation to the all-up mass. On long migratory flights, the bird gets progressively lighter as it uses up fat, and the effective lift:drag ratio increases as the body slims down, so reducing its drag. For both reasons, the fat consumed per metre flown decreases progressively during the flight. To use a subsidiary program that takes account of these effects, select Number 15 from the main menu of Program 1 (not available in Program 1A). It prompts you to supply an estimate of the mean effective lift:drag ratio during the flight, and the mass of fat available for consumption. To get the former, you have to run the standard program (Number 14) twice, once with the full take-off fat load entered as 'Payload', and again with zero payload. The mean of the two estimates for the maximum effective lift:drag ratio is then entered in the range program, and the range (Y) is found from equation (3.15).

Transferring the programs to other computers

The program disc supplied with the book is in MS-DOS format, for IBM-compatible computers. Many computers use operating systems that are not compatible with MS-DOS. Do not despair if you have a disc that is in the wrong format for your computer. The programs are recorded on the disc in the form of ASCII files, that is, ordinary text files, as used for documents. Routines may be available to reformat these files under your computer's operating system. Failing that, the files can be transferred from nearly any computer to nearly any other via an RS-232 link. If both computers have modems, this can be done over a telephone line. Once you have a copy of the program files in the right computer, you should be able to load them into BASIC as above.

Entering the programs from the listings

Full listings of all three programs are in Appendix 1, together with an amendment listing for changing Program 1 into Program 1A. If less arduous methods fail, the user can enter the programs from these listings, although this is not a task to be undertaken lightly. Before attempting it, the reader is urged to heed the notes and advice in the rest of this chapter.

Compatibility with other versions of BASIC

These programs were written in a version of Microsoft BASIC supplied with the Commodore Amiga computer, and were subsequently found to run without modification on various computers running MS-DOS versions of Microsoft BASIC, including the widely used ABASIC and GWBASIC. Every effort was made to use only those features of BASIC that have not been too recently introduced, and are not specific to particular computers. However, some editing may be required to get the programs to run on old computers, or under versions of BASIC from sources other than Microsoft. Some potential sources of trouble are noted below.

Program entry

On most computers, there are two ways to enter a BASIC program, either directly into the BASIC editor, or in the form of a text file, created by a word processor. The text file can then be loaded into BASIC, and debugged using the BASIC editor. Some word processors produce files that cannot be read by a BASIC interpreter, because the output file contains embedded codes that are not visible on the screen. If a 'non-document' or 'text-only' mode is provided for saving files, this will usually do the trick. The same problem may occur in reverse, when using a BASIC editor, and saving a program to disc. Unless instructed otherwise, BASIC will save the program in an encoded form that is not intelligible to word processors. Usually, you can defeat this by selecting an option on the SAVE command, which saves the program as an ASCII file, that is, an ordinary text file. If you cannot find the BASIC manual for your computer, try this syntax:

```
SAVE "Prog_1.BAS",A
```

When to press 'Return'

Each program line should be entered without a break, pressing 'Return' only at the end of the logical line, even though it may overflow on to a second line as printed here. Every BASIC line must be terminated by pressing 'Return', *including the last line in the program.* Also, if a line happens to fill the exact width of the screen, you still have to press 'Return', even though the computer will move to the beginning of the next line of its own accord. This results in a blank line on the screen.

Comment lines

Lines beginning with REM can be omitted. They are included only to make the program more intelligible to the reader, and are ignored by the computer. In Line 50 and subsequently, comments are included on the same lines as program statements. Everything following the apostrophe sign (′) is comment, and is ignored by the computer. This convention may not work in some versions of BASIC. In this case, the comment can either be omitted, or put on a separately numbered line, beginning with REM. Use a new line number, in between two existing lines.

Spaces and punctuation

Spaces are needed where shown (e.g. after line numbers and reserved words), and are generally optional otherwise. When using a word processor, a space may be inserted to invoke the word-wrap function, and break an over-long line in a convenient place. It is vital to enter punctuation exactly as shown, for instance colons (:) and semicolons (;) have entirely different meanings in BASIC, and must on no account be confused. Brackets must be exactly as shown. The exclamation mark (!) after long integers is inserted automatically by Amiga BASIC, but may not be required in some other versions. Blank lines are included for clarity only, and may be omitted.

Line numbers

These programs have been written with line numbers, for compatibility with versions of BASIC that require them. The line numbers are part of the structure of the programs, and must be entered exactly as shown, even if the programs are to be run on a version of BASIC in which line numbers are optional. The programs can be edited so as to dispense with the line numbers, but this is not a trivial task.

Syntax problems

Some early versions of BASIC may not be able to cope with the IF–THEN–ELSE syntax. In this case, all those lines that contain 'ELSE' will need to be located, preferably by using the 'Search' function on a word processor, and modified. A construction that avoids the use of ELSE is illustrated by the drag coefficient calculation in Lines 380–410 of Programs 1 and 1A. The more advanced syntax structures provided in Amiga BASIC have been avoided in these programs, so as not to cause difficulties on other computers.

In Microsoft BASIC, the PRINT statement sends output to the screen, whereas LPRINT sends it to the printer. Some other versions of BASIC use PRINT for both purposes, combined with a separate statement which directs output to one channel or the other until further notice.

Variable names

Variable names have mostly been chosen to resemble the usage of Pennycuick (1975). Greek letters have been rendered phonetically, e.g. rho and eta for ρ and η. Lengthy variable names are allowed in Amiga BASIC, and have been used here, but in some versions of BASIC, variable names are restricted to two characters. More insidiously, some computers will accept variable names such as Vmp and Vmr, but fail to distinguish between them, because they only use the first two characters. It is best to check whether the computer has this problem, by typing the following from the keyboard, with BASIC in direct mode:

```
Vmp=2
Vmr=3
PRINT Vmp,Vmr
```

If the computer can distinguish between these variable names, it will print a '2', then a '3'. If not, it will print '3' and '3', because that is the latest value you assigned to the variable Vm. In this case, the following changes to variable names are recommended: change Vmp to vx, Vmr to vy, Pmin to px, Pmr to py, and Ppro to pz. In Program 2, change Cdw to Cw. All other variable names have been chosen so that they can be truncated to their first two characters, without causing any confusion. If the program is being entered on a word processor, these changes should be made by using the 'search and replace' function. Otherwise a careful search will have to be made to find all occurrences of the offending variables.

Subscripts are not easily handled in BASIC, so subscripted variables such as V_{mp} are rendered with ordinary lower case letters instead of subscripts (Vmp). This convention is also used in the output tables. Capital letters are used for BASIC reserved words, and variable names are in lower case, or mixtures of upper and lower case. Amiga BASIC does not distinguish between upper and lower case letters in variable names, and will treat Vmp, VMP, or vmp as the same variable. Sometimes it automatically changes the case of a letter here and there, following some inscrutable rule of its own. Consequently, upper and lower case letters have been used indiscriminately (and possibly inconsistently) in these programs. Amiga BASIC is indifferent to this, but it may be prudent to make a simple test in direct mode (as above), to check that this is also true of the user's version of BASIC.

All versions of BASIC forbid the use of BASIC reserved words as variable names. One can never be sure, however, that some future version of the language will not contain a newly invented function or command, which happens to be called Vmr, or some other name that is used in the program. Also, some versions of BASIC for IBM-compatible computers forbid variable names that are the same as the names of commands in the MS-DOS operating system. Again, it is not possible to anticipate the introduction of new commands, which could provoke this problem. Users who encounter problems of this type can either change the offending variable names, or try an older version of BASIC.

Testing and debugging

Test of menu functions

Having loaded a newly created copy of Program 1 or 1A, the menu functions should first be tested. When the main menu comes up, choose each option in turn that invites you to change the value of a variable. For instance, if you choose Number 1, you will be prompted 'New species name?', and having supplied this you will also be prompted for a new empty mass, payload mass, and wing span. The menu will then reappear, showing the amended values you have just entered. Try this with all the variables on the list, and make sure that your amended values are duly displayed when the menu comes back.

Program test

If the menu functions work correctly, you can now test the program by running the examples for which full output tables are reproduced in

Appendix 1. Instructions for running the examples, and further information about them, are in Chapter 8. The output should be exactly as shown in Tables A1.1 to A1.9. The layout should be exactly the same, and the numbers should agree to the last decimal place. On no account should results calculated with newly entered programs be used for publication, unless the test examples have been successfully run, with no errors or discrepancies.

As it is nearly impossible to enter a program of this length without making a single typing error, it has to be anticipated that the program will not run correctly at first. Some patience and perseverence will most likely be needed to get it working. Users who are not in the habit of writing their own programs often become convinced that the computer is broken, but usually its irrational behaviour is eventually traced to typing errors in the program. Even apparently trivial mistakes can cause mystifying malfunctions. If the results do not match the specimen output, there is an error, which must be found and corrected before the program is used in earnest. Errors fall into two main categories, as follows:

Syntax errors A typing error that causes a statement not to 'make sense', as understood by BASIC, will cause program execution to be aborted. BASIC returns to direct mode, after printing a message saying 'Syntax Error', and indicating the line number in which the problem occurred. Often the cause is a missing bracket (or one too many), or a misspelled word. Common causes of these errors in PRINT and LPRINT statements are semicolons missing or confused with colons, or quotes (") missing, or in the wrong place. The FNr function, which rounds numbers to a fixed number of significant digits, is not part of BASIC, but is a 'user-defined' function. The definition is in Line 100, in all the programs. Every time the FNr function is used, the computer runs Line 100. If there is an error in Line 100, this causes a message saying 'Illegal Function Call', but the message indicates the line where the function was *called*. A similar message can occur if the number on which the function operates (its argument) somehow manages to be zero or negative. This can be caused in various devious ways, by errors in places nowhere near the apparent seat of the trouble. For instance, if you get this message when the computer is trying to print out Vmr, after it has already printed out Vmp successfully, try printing Vmp and Vmr from the keyboard. If you do this immediately after program execution fails, you can find out what the values of the variables were when it failed.

Run-time errors If a statement is syntactically correct, it will be executed, but may still produce a wrong result. For example, if an intricate statement like Line 850 is typed with the correct number of brackets, it will be executed, even though one of them is in the wrong place. This causes a 'run-time error', that is, the program runs, but the results are not what you expect. Finding a run-time error entails identifying the point in the program sequence at which the calculation went off the rails. Misspelling of a variable name is a common cause of run-time errors. For example, suppose that in Program 1 or 1A you put 'Ppor' instead of 'Ppar' in Line 990, which finds the total power by adding up Ppar and several other components of power. As far as BASIC is concerned, Ppor is a valid variable name, to which no value has been assigned, so it is assigned the default value, which is zero. Line 990 will then use zero instead of the real value of Ppar, when it adds up the power. This does not stop the calculation, but the powers all come out too low, especially at high speeds. As a diagnostic aid, you can insert an extra print statement, giving it a line number between two existing lines, to find out whether the values of certain variables are what you expect them to be at a particular point in the program. After a program run has been completed or aborted, you can also execute PRINT statements from the keyboard as indicated above, to find out the values of variables on exit from the program.

Rounding of results

All the programs round results to three significant digits, using the FNr function in Line 100. This works perfectly on Amiga and Macintosh computers, but some IBM-compatible computers occasionally print a number ending in a string of nines, or in a string of zeros ending with a one. This seems to be caused by lack of precision in the computer's arithmetic routines, not by any defect in the BASIC. The only known cure at present is to get a better computer.

5. The reality behind the power calculations

Introduction

The discussion in Chapter 3 represents the flying bird in terms of a number of simplified entities, beginning with the body, which creates drag according to a specified law, and the wing disc, which accelerates air downwards in an exceedingly clean and tidy manner. Of course, these abstractions represent (at best) a simplified account of what happens when a real bird proceeds on some errand through the atmosphere. It is temptingly easy to make the theory much more elaborate, but this is not a path to be followed incautiously. As a general rule, the best theory is the simplest that predicts the results within acceptable limits of accuracy. It is usually not justifiable to add complications to a theory, unless it can be shown that a more accurate result is obtained by doing so. At present there is not a great deal of scope for adding elaborations to the power curve theory of Chapter 3, for two reasons. First, there is very little information from direct observation about the physical processes that go on in the air around a flying bird. Second, there is almost no reliable information about the amount of power that birds actually consume in flight, notwithstanding a large number of reported measurements of 'flight metabolism' (more about this in Chapter 8). The objective of this chapter is to explain in more detail the physical processes underlying the abstractions of Chapter 3, so that the reader can appreciate the nature of the simplifications that have been made, and see in qualitative terms where errors are likely to arise.

Reynolds number and the concept of scale

As one proceeds from insects to birds, and on to airliners, both the size and the speed progressively increase. Both trends contribute to an increase of 'scale', as the term is understood in engineering. The properties of the fluid also help to define the scale. The pattern of flow around bodies or wings changes progressively as one moves from a small to a large scale. The *Reynolds number* (*Re*), is used as an index of scale. It is designed to take account of size, speed, and the properties of the fluid in a single number, and is defined as

$$Re = Vl/\nu, \quad (5.1)$$

where V is the airspeed, l is some representative length, and ν is the *kinematic viscosity* of the air, that is, the ratio of the viscosity to the density. The dimensions of kinematic viscosity are L^2/T, and the SI units are square metres per second. Values for air at various heights are listed in Appendix 4.

The Reynolds number is dimensionless, meaning that no units are required to express it, and also that the same number results whether you express V, l, and ν in SI or foot-pound units, provided that the units all belong to the same family. However, the value of the Reynolds number is to some extent arbitrary, depending on the practical definition chosen for the 'reference length' (l). For a streamlined body, it is most appropriate to define l as the total length of the body, along the direction of flow. However, streamlined bodies used for wind tunnel testing are usually circular in cross section, and it is traditional to use the diameter at the widest point, rather than the length, as the 'reference length'. If you base your Reynolds numbers on the length, while your neighbour bases his on the diameter, your Reynolds numbers will be several times larger than his, even though the conditions of flow are the same for both. It is essential to know exactly what length a quoted value for the Reynolds number is 'based on'. Reynolds numbers quoted for whole wings are usually based on the mean chord (see Chapter 2). However, in sailplane engineering, where this is of critical importance, authors often base their Reynolds numbers on the local chord, and regard the Reynolds number as being lower near the tip of a tapered wing than near the root. In the case of bird wings with splayed primaries, it would be appropriate to quote a lower Reynolds number for an individual primary than for the wing as a whole. Some estimate of Reynolds number should always be given in connection with observations of bird flight, as this will give the informed reader an immediate indication of the type of flow regime likely to be encountered. Often two Reynolds numbers are given, one for the body, based on the diameter at the widest point, and the other for the wing, based on the mean chord.

Similarity criteria

The original use of Reynolds number by engineers was as a 'similarity criterion', for determining whether the flow around a small-scale model, being tested in a wind tunnel, is similar to that around the full-sized aircraft in free flight. One of the requirements for this is that the Reynolds number should be the same for the model and the full-sized aircraft. For instance, a half-scale model would have to be tested at twice the full-size flight speed to get the same Reynolds number. Alternatively, engineers sometimes test models of aircraft parts in water instead of air, because the kinematic viscosity of water is only about one-tenth that of air at sea level, and therefore, from equation (5.1), a similar Reynolds number can be achieved on a small-scale model without requiring an excessively high speed. Animals of different size never go about their errands at similar Reynolds numbers, because there is a general tendency for larger animals to fly at higher speeds than smaller ones. Since both V and l in equation (5.1) are larger for a large animal than for a small one, the larger animal inevitably flies at a higher Reynolds number. It follows that the pattern of flow is expected, in general, to be different in large and small animals, even though the body shape may be much the same.

Some criterion is still needed for deciding how to make comparisons under 'similar' conditions. Often, the flight of different-sized animals is compared when each is flying at its minimum power speed, or at its maximum range speed. In this case the Reynolds number increases progressively (and strongly) with body size, with accompanying changes in the pattern of flow.

Inertial and viscous forces

Reynolds number can also be understood as an expression of the ratio of inertial forces to viscous forces acting on a body. If you imagine a little cube of fluid making its way over a curved surface as in Fig. 5.1, two sorts of forces are needed to accelerate and decelerate it, and deflect it around the curves. The inertial forces are proportional to the mass of the cube, whereas the viscous forces are proportional to its surface area. If you scale the whole drawing up, the ratio of the cube's mass to its surface area increases, and therefore the inertial forces needed to propel it along the same path become more prominent in relation to the viscous forces. The same effect is produced by making the fluid more dense or less viscous. On

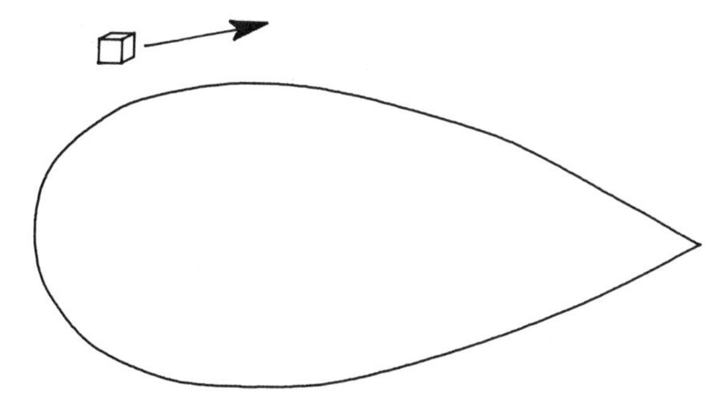

Fig. 5.1. *An imaginary cube of air following the surface of a streamlined body.*

the very small scale of microbes moving about in body fluids, viscous forces predominate, and inertial forces are insignificant. In the large-scale world of airliners, ships, and whales, the opposite is the case. The forces are predominantly inertial, and viscous effects only modify the way in which the inertial forces act.

Readers who require more information about scale and Reynolds number will find an excellent discussion of the subject, written for biologists, in Vogel (1981).

Drag and drag coefficient

The air flowing around a streamlined body, as in Fig. 5.2, causes drag for two quite different reasons. The two types of drag are known as *pressure drag* and *skin friction*. Pressure drag is due to the fact that the air pressure varies over the surface, being highest at the front and back ends, and lowest where the air is flowing fastest around the central bulge. The pressure at each point pushes at right angles to the surface. The pressure acting on each element of surface at the front end of the body has a component pushing backwards (i.e. adding to the drag), while the pressure acting on each element at the rear end of the body contributes a forward component. The sum of all the backward components of force is somewhat greater than the sum of all the forward components, and the difference between them is the pressure drag. If the pattern of flow were to stay the same as the speed is varied, the front-to-back pressure difference would be proportional to the dynamic pressure, and hence the pressure drag would vary with the

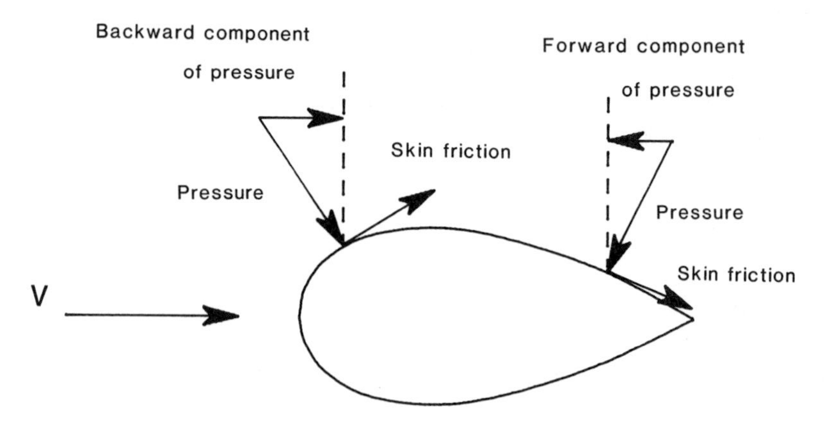

Fig. 5.2. *The sum of all the backward components of pressure is greater than the sum of all the forward components, and the difference is the pressure drag. All the components of skin friction drag act backwards.*

square of the speed, i.e.

$$D = \tfrac{1}{2}\rho S C_D V^2, \quad (5.2)$$

where ρ is the air density, S is the cross-sectional area of the body, V is the speed, and C_D is a dimensionless *drag coefficient*. Equation (5.2) can be rearranged to find the drag coefficient from observed values of the drag:

$$C_D = 2D/\rho S V^2. \quad (5.3)$$

If the drag is indeed proportional to the square of the speed, then the drag coefficient will be independent of speed, since both the numerator and the denominator of equation (5.3) vary with the speed squared.

Skin friction is due to the stickiness of the air. As it slides over the surface of the body, it exerts a force *parallel* to the surface of each element of area. All of these elements of force have a backward component, and the sum of all the backward components is the skin friction. The magnitude of the skin friction force is directly proportional to the speed (not to the speed squared). Thus a drag coefficient for skin friction, if defined according to equation (5.3), would vary inversely with the speed.

Pressure drag (being due to inertial forces) predominates at high Reynolds numbers, while skin friction drag predominates at low Reynolds numbers. The use of a constant drag coefficient, as in equation (5.2), implies that:

(1) the body drag is assumed to be mostly due to pressure drag, with skin friction making only a minor contribution;

(2) as speed is varied, the pattern of flow does not systematically change in a way that would affect the pressure drag coefficient.

These two assumptions are reasonably good for full-sized aircraft. The first assumption is not good for insects, while birds, bats, and model aircraft inhabit an intermediate zone in which the second assumption has to be treated with caution.

Flow around a sphere at different Reynolds numbers

The pattern of flow around a sphere illustrates a number of principles, which also apply in modified form to the flow around bodies of more useful shapes. At very low Reynolds numbers the flow is symmetrical upstream and downstream as in Fig. 5.3. The pressure varies little over the surface, and the drag, for practical purposes, is entirely due to the tangential force caused by the fluid sliding slowly past. A non-viscous fluid (if there were any such substance), would not exert any drag at all under these circumstances. Actually a non-viscous fluid would not exert any drag at higher Reynolds numbers either. If the speed is sufficient for inertial forces to be appreciable, then the fluid pressure at the 'stagnation point', in the

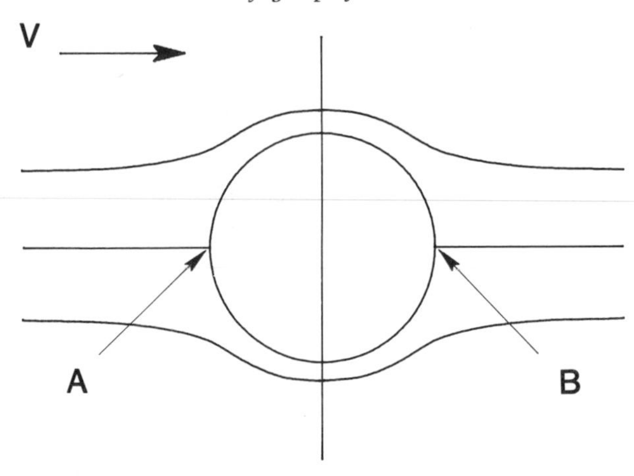

Fig. 5.3. *In a fluid with zero viscosity, the flow round a sphere would be symmetrical, with equal pressure on the upstream and downstream sides, and therefore no pressure drag. The pressure* (p) *at both the upstream and downstream stagnation points (A and B), is higher than the pressure* (p_0) *in the undisturbed stream, by an amount equal to the dynamic pressure* ($\frac{1}{2}\rho V^2$).

middle of the upstream side of the sphere, is increased above the free-stream static pressure by an amount

$$p = \tfrac{1}{2}\rho V^2, \tag{5.4}$$

that is, by the dynamic pressure. A pressure gradient along the surface causes the fluid to accelerate until it reaches the widest point, where the pressure is less than ambient. The pressure then increases again until it once again reaches $\frac{1}{2}\rho V^2$ at a second stagnation point, in the middle of the downstream side of the sphere. The pressure on the back side of the sphere is the same as that on the front, so they balance and there is no pressure drag. Since the fluid is non-viscous by definition, there is no skin friction either, and no total drag. This theoretical result is known after its discoverer as 'd'Alembert's paradox', because everyone knows that real spheres do experience drag. In the real world, at some moderate Reynolds number, the flow around the downstream side of the sphere is not quite a mirror image of that on the upstream side. The stagnation pressure on the upstream side is indeed $\frac{1}{2}\rho V^2$, as it is supposed to be, but that on the downstream side is less. The pressure difference between the upstream and downstream sides results in pressure drag. This effect is due to inertial forces, but the reason it occurs is that viscous forces cause the pattern of flow to deviate from the symmetrical arrangement shown in Fig. 5.3. At high Reynolds numbers, viscous effects, though relatively small in themselves, lead indirectly to much larger inertial forces, which appear as pressure drag.

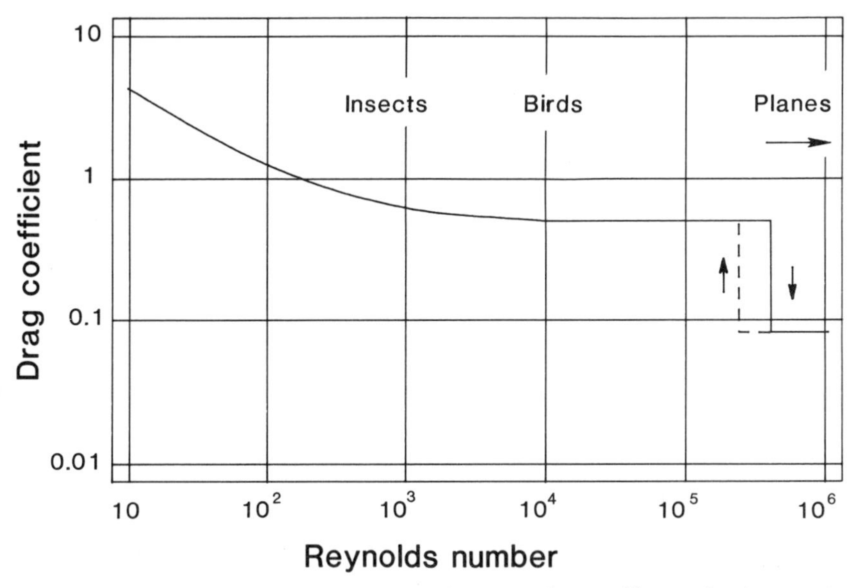

Fig. 5.4. *Drag coefficient of a sphere, as a function of Reynolds number (composite graph from information in Vogel 1981 and Schmitz 1960).*

Drag coefficient of a sphere

Figure 5.4 is a double-logarithmic graph of the drag coefficient of a sphere, plotted against Reynolds number, over a range extending from the scale of microbes ($Re = 10$) to the lower end of the aircraft range ($Re = 10^6$). At very low Reynolds numbers the drag is proportional to the speed, and the drag coefficient is inversely proportional to the Reynolds number, as noted above. The curve therefore comes down initially at a slope of -1, but gradually levels out as inertial forces become more prominent in relation to viscous forces. Above about $Re = 20\,000$, the drag is predominantly pressure drag, and the drag coefficient levels off at about 0.48. It holds this value until about Re = 405 000, and then suddenly makes a large downward jump to a value of about 0.08. This value remains steady, up into the aircraft range of Reynolds numbers (1 million upwards). The downward jump in the drag coefficient at $Re = 405\,000$ is due to a change in the character of the flow in the *boundary layer*, that is, the layer of air near the surface, in which there is a velocity gradient from the free stream to the lowest layer of air molecules, which are stuck to the surface and do not move at all. The boundary layer is where the frictional forces are generated, because it consists of layers of air that are sliding over one another. At low Reynolds numbers, the flow in the boundary layer is said to be *laminar*, meaning that layers of air slide smoothly over one another on parallel paths.

Boundary layer separation

On the upstream face of the sphere, the air in the lower layers moves in the direction of flow for two reasons. First, friction exerted by the layers above pulls the lower layers along. Second, the static pressure is highest at the stagnation point in the middle of the upstream side of the sphere, and lowest near the mid-point. There is therefore a pressure gradient, tending to accelerate air near the surface in the downstream direction. Beyond the minimum pressure point, the static pressure starts to rise again, so that the gradient of static pressure is reversed, and tends to push the lowest layers of air against the direction of flow. The flow reversal near the surface undercuts the boundary layer, and causes it to separate from the surface (Fig. 5.5a). Instead of following the curve of the sphere round to the downstream side, as in Fig. 5.3, the flow breaks away, leaving a large area of chaotic flow on the downstream side. The pressure in this *chaotic wake* is near ambient, whereas that at the upstream stagnation point is equal to the dynamic pressure. This large front-to-back pressure difference results in a large amount of pressure drag, represented by the drag coefficient of 0.48 along the level part of the curve in Fig. 5.4.

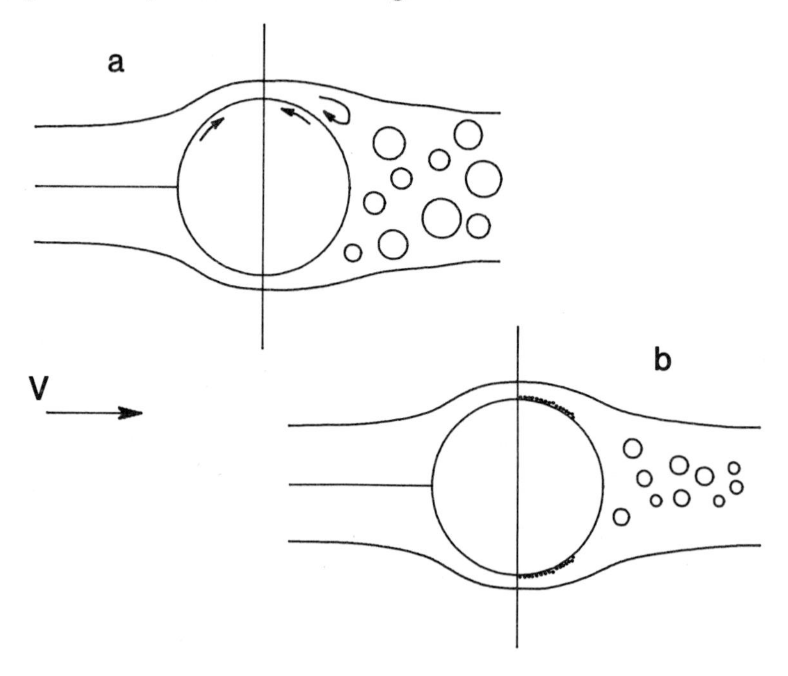

Fig. 5.5. *Flow around a sphere at Reynolds numbers (a) below and (b) above the critical value at which the flow in the boundary layer changes from laminar to turbulent.*

The downward step in the drag coefficient curve is due to the flow in the boundary layer changing abruptly from laminar to *turbulent*. This boundary-layer turbulence is on a microscopic scale, and should not be confused with the larger-scale turbulence in the chaotic wake. Air molecules shuttle up and down through the boundary layer, carrying momentum between the free stream and the surface. Because of this, the lowest layers of air are much more strongly dragged along in the direction of flow than they are when the boundary layer is laminar. The undercutting of the boundary layer downstream of the minimum pressure point, which causes separation of the laminar boundary layer, no longer occurs. The turbulent boundary layer remains attached to the surface, much further round towards the downstream side. The width of the chaotic wake is greatly reduced, and so is the pressure drag caused by it (Fig. 5.5b). In the terminology of Schmitz (1960), the flow pattern with a separated, laminar boundary layer, wide wake, and high drag coefficient is said to be 'subcritical'. When turbulence sets in, the wake contracts, and the drag coefficient decreases, the flow is said to be 'supercritical'.

It may be noted that the transition from laminar to turbulent flow in the boundary layer causes the skin friction to *increase*. However, in the case of a sphere at intermediate Reynolds numbers, the drag is mostly pressure drag. The small increase in skin friction is masked by the large reduction in pressure drag, which results when the onset of turbulence in the boundary layer causes the flow to become attached to the surface. If a sphere is tested in a special wind tunnel that has been designed to create a very smooth flow of air, the transition from laminar to turbulent flow occurs spontaneously as the wind speed is gradually increased, at about $Re = 405\,000$. If the wind speed is then gradually decreased, there may be a hysteresis effect, the flow remaining turbulent down to a lower Reynolds number, before spontaneously changing back to laminar. The transition to turbulent flow can also be stimulated at a lower Reynolds number than 405 000 by small amounts of turbulence in the air stream, or by a small, sharp projection fixed to the surface of the sphere upstream of the minimum pressure point. Aeromodellers often fix 'turbulator strips' to the surface of bodies and wings, to make sure that the boundary layer is turbulent ahead of the minimum pressure point, thus keeping the flow attached to the surface.

Drag of bird bodies

Wind tunnel tests on bird bodies are readily understood in terms of the behaviour of spheres as illustrated in Fig. 5.5. By convention, the Reynolds number is based on the diameter of a circle, whose area is the same as the cross-sectional area of the body at the widest point (or frontal area). This is directly comparable to Reynolds numbers for a sphere, based on the diameter of the sphere. Various authors have made drag measurements on frozen bird bodies, from which the wings had been removed, mounted in a wind tunnel. Early measurements on a pigeon at $Re = 41\,000$, and a Rüppell's griffon vulture at $Re = 140\,000$ both gave estimates of $C_D = 0.43$ (Pennycuick 1971), which is only slightly less than the drag coefficient of a sphere in the subcritical range of Reynolds numbers. Prior (1984) tested a number of waterfowl bodies, including various ducks and swans, at Reynolds numbers from 65 000 to 300 000. He observed drag coefficients up to about 0.4 in the lower part of this range, decreasing and levelling off at about 0.20 above about $Re = 200\,000$. Prior also used smoke filaments to observe the contraction of the wake associated with this reduction in drag coefficient. Pennycuick, Obrecht, and Fuller (1988) tested some waterfowl and raptor bodies at Reynolds numbers from 145 000 to 462 000 (overlapping the higher part of Prior's range) and observed somewhat higher drag coefficients, averaging about 0.32 in the supercritical range. Tucker (1973) reported higher drag coefficients for smaller birds.

All of these studies suffered from technical problems, which makes it difficult to be sure how representative the drag measurements are, as compared to real bird bodies, with the wings in place and flapping. One feature they have in common is that the measured drag coefficients were high, indicating that the chaotic wake was wide, and the boundary layer detached over the posterior part of the body. This conclusion is reinforced by Prior's smoke-filament experiments, and by observations of visible agitation of the feathers at the posterior end. It seems clear that medium-sized and large birds fly in just that range of Reynolds number where aeromodellers find it helpful to use turbulator strips to keep the flow attached. One might expect to see sharp projections in the shoulder region, serving to trigger turbulence in the boundary layer. No such projections are seen in a bird body, but it is possible that the wing leading edge serves this function in an intact bird. Be that as it may, it is difficult to believe that intact flying birds have quite so much body drag as the wind tunnel results suggest, especially as close-up films of free-flying birds do not show so much disturbance of the feathers around the posterior part of the body as is seen in wind tunnel tests. It is possible that the feathers themselves operate in some way to keep the boundary layer attached in the living bird. In the experiments of Pennycuick, Obrecht, and Fuller (1988) it was noted that the feathers became fluffed out when the bodies were tested at high speeds, and this seemed to increase the drag. When a snow goose body was coated with hair spray, to glue the feathers down, the drag decreased by about 15 per cent. In a living bird the feathers can be controlled by pteromotor muscles, and it may be that the bird is able to exert some control over its boundary layer by erecting or flattening the contour feathers.

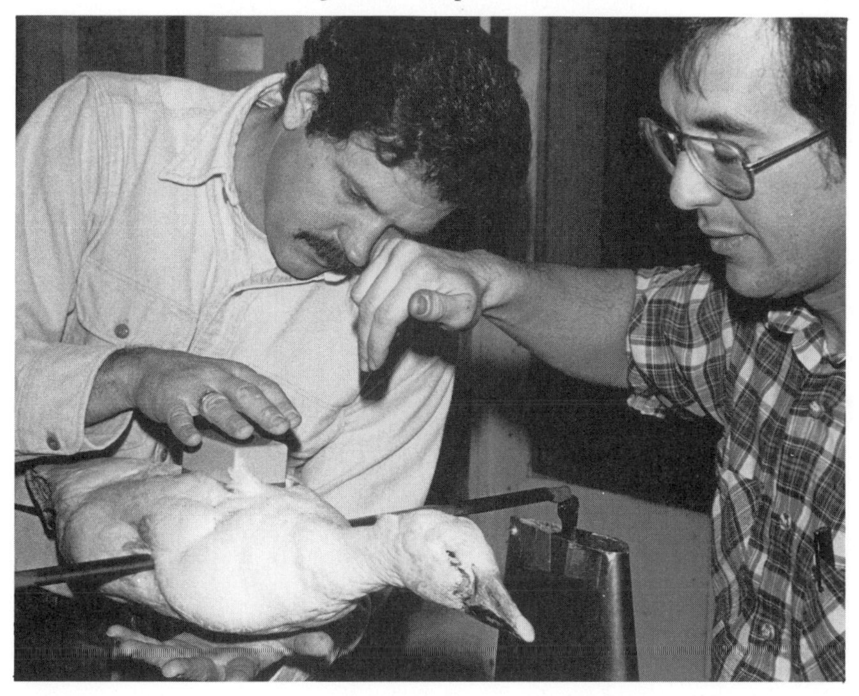

Fig. 5.6. *A snow goose body being set up for a drag measurement in a wind tunnel. In this test, the body was fitted with a large, dummy radio transmitter, to determine the effect on the drag. University of Maryland.*

For the time being, the high values for body drag obtained from the experiments have been incorporated in the computer programs given in this book. The programs make a rough estimate of the body's Reynolds number in cruising flight, and then take the drag coefficient from the graph of Fig. 5.7. This assumes that the drag coefficient is 0.4 in the subcritical region up to $Re = 50\,000$, then declines to a value of 0.25 at $Re = 200\,000$, and remains constant at higher Reynolds numbers. The reader will appreciate that this is a somewhat rough and ready compromise, derived from a number of experimental results which do not agree well with each other. The experimental data, such as they are, refer mainly to medium-sized and large birds, and there is little direct evidence concerning the body drag of passerines or hummingbirds. All of the experimental data are under suspicion of being biased upwards, because of more extensive separation of the boundary layer under experimental conditions, than occurs in free flight. If this suspicion proves to be justified, there is a real possibility that future research will reveal that the current drag coefficient estimates need to be revised downwards. In the meantime, the reader can, of course, change

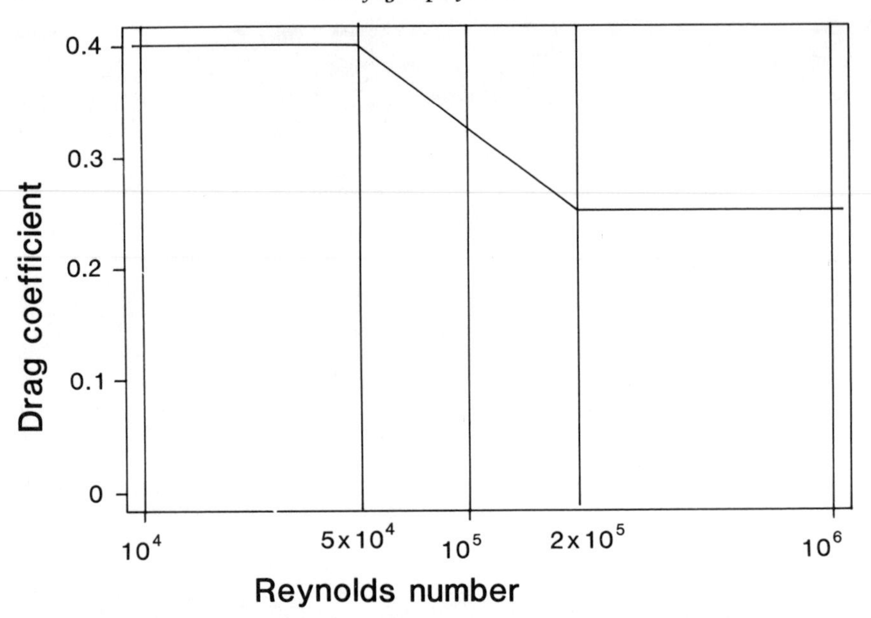

Fig. 5.7 Rule for estimating the drag coefficient of bird bodies, as a function of Reynolds number at V_{mr}, *based on the diameter of the body. After Pennycuick, Obrecht, and Fuller (1988).*

the value of the body drag coefficient calculated in any of the programs, by inserting a new value from the main menu.

Profile power and the flow around wings

The flow around a wing is more complicated than that around a non-lifting body, because the function of a wing is to deflect the air, that is, to give it some momentum in a direction perpendicular to the incident air flow. The force resulting from this is the *lift*, which acts, by definition, at right angles to the direction of the incident air flow. The magnitude of the lift force is equal to the rate at which transverse momentum is applied to the air. The lift force does not necessarily act upwards. If the bird happens to be executing a vertical dive or climb, the lift on its wings (if any) would act horizontally. In a flapping wing, different parts of the wing 'see' air approaching from different directions, and in this case the local contribution to the lift is perpendicular to the direction from which the air appears to be approaching each part of the wing.

The deflection of the incident air out of its original path is accomplished mainly by the area of reduced pressure above the wing, which sucks the

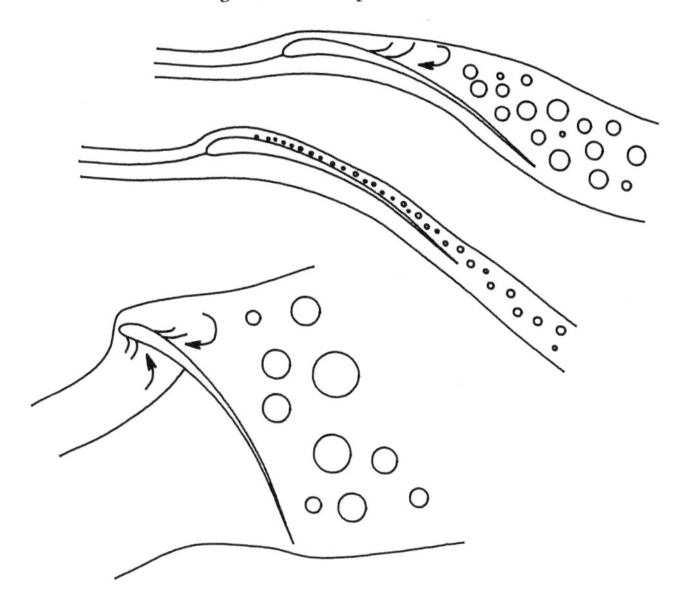

Fig. 5.8. *At low Reynolds numbers (top), there is tendency for the boundary layer to separate from the upper surface of birds' wings. Reversed flow lifts feathers as in Fig. 5.9, and there is a wide chaotic wake. At higher Reynolds numbers (middle) the formation of small-scale turbulence in the boundary layer reduces the tendency for separation, and narrows the chaotic wake. At very high angles of attack, as in flapping flight at low speeds (bottom), the stagnation point moves to the underside of the wing. Raised feathers sometimes reveal reversed flow on both upper and lower surfaces simultaneously, as in Fig. 5.10.*

air around the curve. As with a non-lifting body, the pressure reaches a minimum at some point along the chord, and thereafter increases again, reaching ambient at least by the time the air reaches the trailing edge of the wing, if not sooner. Just as with a non-lifting body, there is a region of reversed pressure gradient, which tends to cause reversed flow in the boundary layer. The resulting separation of the flow from the upper surface of a wing not only causes a large amount of pressure drag, as in a non-lifting body, but also leads to a loss of lift, because it defeats the wing's primary function of deflecting the airflow.

The pressure drag and the skin friction together constitute the profile drag (D_{pro}) of a wing. As with the drag of a non-lifting body, it can be represented in terms of a non-dimensional *profile drag coefficient*, defined as

$$C_{Dpro} = 2D_{pro}/\rho SV^2, \tag{5.5}$$

where S is the wing area as defined in Chapter 2. The lift generated by the

Fig. 5.9. *A gliding fulmar, with reversed airflow picking up the feathers over the proximal part of the wing. Shetland.*

wing can be represented in a similar way as a non-dimensional *lift coefficient* (C_L), defined as

$$C_L = 2L/\rho SV^2, \tag{5.6}$$

where L is the magnitude of the lift. The lift coefficient is closely related to the angle through which the air is deflected by the wing. If the boundary layer separates, the profile drag coefficient increases because of the increased area of chaotic wake, while the lift coefficient decreases, because the air is deflected through a smaller angle.

As with non-lifting bodies, the boundary layer is more prone to separation at low than at high Reynolds numbers, and laminar boundary layers separate more readily than turbulent ones, but create less skin friction (Fig. 5.8). Sailplane designers go to great lengths to maintain large areas of laminar flow on their wings in order to minimize skin friction, but aeromodellers cannot do this, because laminar boundary layers separate too easily at the lower Reynolds numbers. Aeromodellers take just the opposite measures to those favoured by their full-size counterparts. For example it is a common practice to glue a thread along the upper surface

Fig. 5.10. *A puffin flapping hard, prior to landing, with feathers lifting on both the upper and lower surfaces of the wing, as in Fig. 5.8. Shetland.*

of a model aircraft wing near the leading edge, in order to trigger turbulence, and ensure that the boundary layer is fully turbulent all over. The penalty in skin friction has to be accepted, in order to keep the boundary layer attached and generate a sufficient amount of lift.

Bird wings are certainly not immune to extensive separation of the boundary layer. Separation is easily observed because it causes the contour feathers on the upper surface of the wing to lift up and flutter about (Fig. 5.8). More or less any of those birds that habitually soar on cliff faces (gulls, petrels, albatrosses, gannets, auks, vultures, ravens, etc.) can often be seen gliding along with an extensive area of separated flow lifting the feathers on the upper surface of the wing, especially when gliding slowly or descending steeply (Figs. 5.8 and 5.9). This must imply that the profile drag is substantial. The profile drag coefficient is also dependent on the lift coefficient, because the tendency for the boundary layer to separate is stronger at high lift coefficients. Under the extreme conditions seen in low-speed flapping flight, it is actually possible to see separation or reversed flow on both sides of the wing at the same time (Figs. 5.8 and 5.10).

There are no direct measurements of the profile drag of bird wings in the literature. The nearest approach is an early estimate for the profile drag coefficient of a gliding pigeon by Pennycuick (1968), which was obtained indirectly, by first measuring the drag of the whole bird, and then subtracting the parasite drag (from measurements on a wingless body), and the induced drag (calculated). The results indicated that the profile drag coefficient was high, and strongly dependent on the lift coefficient. Both conclusions can readily be understood in terms of the qualitative discussion above. In the present vestigial state of empirical knowledge, the method used in Chapter 3 for calculating profile power as a fixed multiple of the absolute minimum power, is as good as any other, *provided that the limitations of speed are observed.* This rough and ready approximation has some chance of being near the truth for speeds from the minimum power speed to the maximum range speed, but not outside that range. It is unlikely to be correct at very low speeds, where the wing is working at high lift coefficients, and probably with extensive areas of separated flow, neither is there any empirical basis for applying other, more intricate methods to this flight regime. At very high speeds, above the maximum range speed, the boundary layer should remain attached, and the profile drag coefficient might reasonably be expected to remain steady. If any birds exist that can fly horizontally at such high speeds (doubtful), one might expect the curve of profile power to bend upwards above the maximum range speed, and eventually to increase with the cube of the speed, like the parasite power.

Induced power and vortex wakes

Vortex wakes

The induced power calculation of Chapter 3 proceeds by considering how the air behind the bird is accelerated downwards. The 'wake' is represented as a simple tube of air, which acquires downward momentum from the wings at a rate equal to the bird's weight. Real wakes are more complicated, and consist of structures in the air, made up of vortices. A vortex wake, which is highly organized, should not be confused with the chaotic type of wake responsible for the drag of spheres and other bodies (above). The vortex wakes generated by wings impart downward momentum to the air. For level flight it remains true, as assumed in Chapter 3, that the *rate* at which the air acquires downward momentum must equal the bird's weight, and the rate at which work has to be done to create the wake is equal to the induced power.

Three simple rules about vortices

The reader requiring a formal and quantitative account of the mechanics of vortices can find it in Milne-Thomson (1958). The following three 'Rules' about vortices take some liberties with Milne-Thomson's standards of

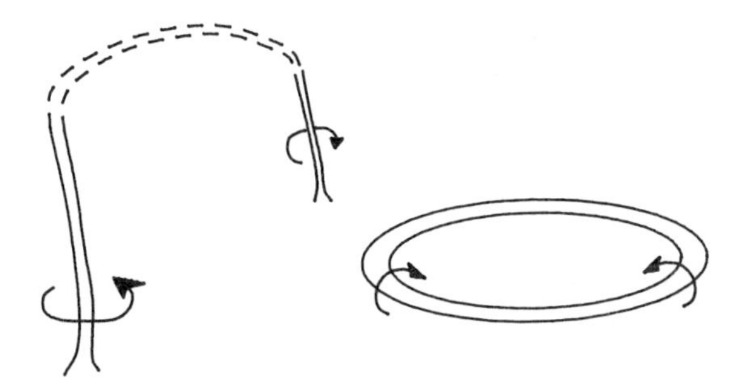

Fig. 5.11. *A vortex is permitted to end by butting against a solid surface (left), but what happens at the top of a dust-devil? No-one knows. It cannot end in the fluid, but might possibly come down again as another dust-devil, rotating the opposite way. A vortex can avoid the need to end in the fluid by joining on to itself to form a vortex ring (right). From Pennycuick (1988).*

mathematical rigour, but will allow the reader to see in a general way how the weight is supported by various kinds of vortex structures, and how the energetic cost of supporting it may be estimated.

Rule 1. A vortex cannot end in the fluid A dust devil is an example of a vortex. It is a rotating column of air, with its lower end in contact with the ground surface. It is permissible for a vortex to end where it butts against a solid surface, but it is far from clear what happens at the top of the dust devil, since a vortex is not permitted to end in the open air. One possibility is that it could curve over to form a horseshoe vortex, with both ends in contact with the ground, rotating in opposite directions (Fig. 5.11, left). It often happens that a vortex avoids the need to end, by curving round and joining on to itself, so forming a ring vortex (Fig. 5.11, right).

Rule 2. Circulation is conserved The 'strength' of a vortex is usually represented in terms of its *circulation*, a quantity with dimensions L^2T^{-1}. It is not necessary for the reader to know how to measure circulation, but only to note that circulation is everywhere constant along the length of a vortex, unless work is done on the system. One large vortex can break up into several smaller ones, or vice versa, but the total circulation remains the same. A related quantity, *vorticity*, is circulation per unit area (dimensions T^{-1}), and shares the same conservation property.

Rule 3. Energy is proportional to circulation and length The amount of energy needed to create a vortex is proportional to its circulation, and also to its length.

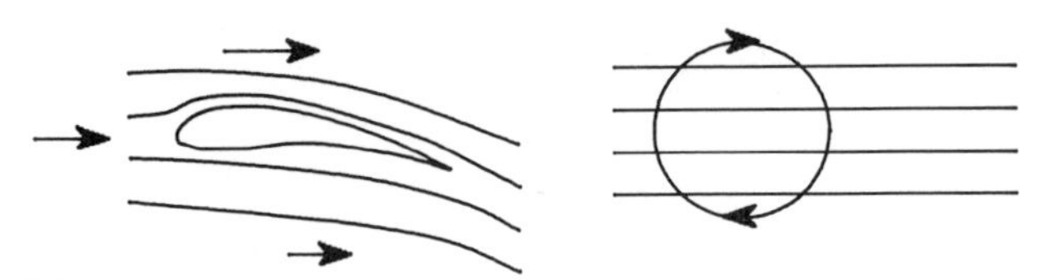

Fig. 5.12. *A wing develops lift by deflecting the airflow downwards (left), which implies that the air above the wing is speeded up relative to the free stream, whereas that below is slowed down. This can also be represented as a* bound vortex, *attached to the wing, superimposed on the undisturbed flow (right). These are two alternative ways of representing the same thing. From Pennycuick (1988).*

Fixed-wing wake

The function of any wing is to deflect the incident airflow, and impart downward momentum to it. In cross-section this implies a pattern of flow like Fig. 5.12 (left), in which the flow is speeded up over the top of the wing, and slowed down underneath. Fig. 5.12 (right) is another way of representing the same flow pattern. The wing is represented by a vortex, called the 'bound vortex', which is superimposed on a steady flow. The speed due to the vortex is added to the free-stream speed above the wing, and subtracted from it below, which produces the same result as Fig. 5.12 (left). On the downstream side of the wing, the flow due to the vortex is downwards, causing the downwash which results in lift. There is a theorem in classical aerodynamics which says that the magnitude of the lift force is proportional to the circulation of the bound vortex, and to the wing span. Thus a given lift force can be produced either with a long wing span and a modest circulation, or with a shorter wing span and a more vigorous circulation (Fig. 5.13). The force to be supported is the same in both cases

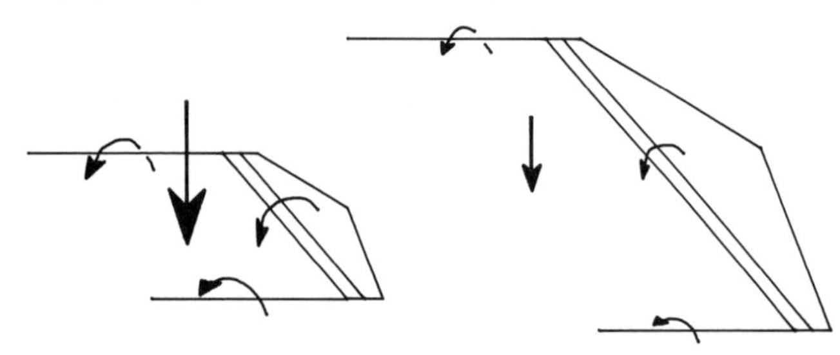

Fig. 5.13. *To support the same amount of weight, a shorter wing must have a stronger bound vortex. The strengths of the trailing wing-tip vortices are therefore also stronger, and the downwash velocity is greater. From Pennycuick (1988).*

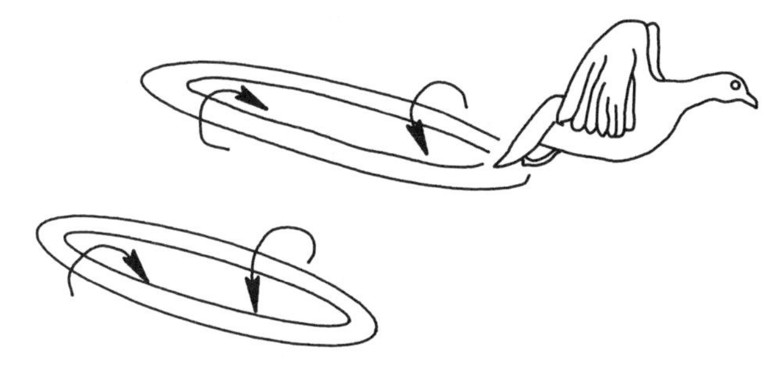

Fig. 5.14. *Single-ring wake in a slow-flying pigeon. One vortex ring is formed at the end of each downstroke, and convects downwards before the next ring is formed. After Spedding, Rayner, and Pennycuick (1984).*

(the weight), but the downwash velocity is higher for the shorter wing, and therefore so is the induced power.

The bound vortex cannot end at the wing tips (Rule 1), so it has to bend around to become a pair of trailing vortices, whose circulation is the same as that of the bound vortex (Rule 2). This combination is often known as a 'horseshoe vortex', but it is not really a horseshoe because the trailing vortices cannot end in the fluid (Rule 1). In the case of a fixed wing aircraft, they are joined together by a 'starting vortex', which is left on the runway at the point where the circulation first developed on the wing. When the aircraft lands, and the load on the wing is relieved, the trailing vortices are joined by a 'stopping vortex' at the point where they end on the runway. Thus the wake caused by one complete cycle of loading and unloading the wing, is actually a very elongated vortex ring.

Wakes in slow flapping flight

Birds in slow flight also apparently reduce the lift to zero during the upstroke of each wingbeat, and therefore create one vortex ring for each wingbeat (Fig. 5.14). The vortex ring has a circulation, and a length (around the ring), and therefore the amount of energy needed to create it can be calculated (Rule 3). The ring also has a property known as 'impulse', which is equivalent to momentum, and this has to be sufficient to support the weight for the period of one wingbeat. Rayner (1979) showed how the induced power can be calculated from the geometry of the ring, which is itself predicted from the bird's wing span and the amount of impulse required. Spedding (1981) and Spedding, Rayner, and Pennycuick (1984) showed that a slow-flying pigeon did indeed create vortex rings in agreement (more or less) with Rayner's predictions.

A bird that flies with a single-ring wake has to be able to fold its wing

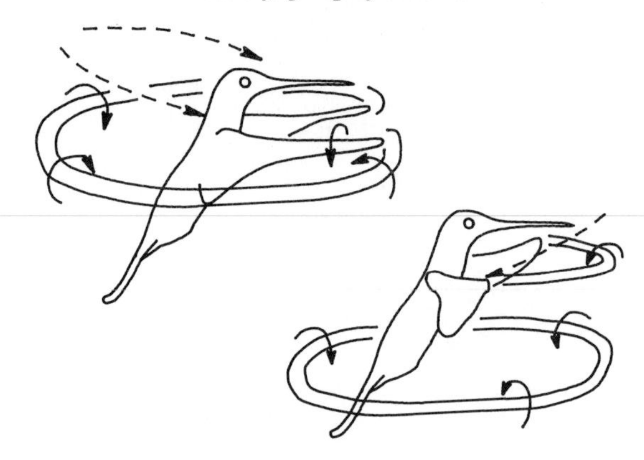

Fig. 5.15. *Double-ring wake in a hovering hummingbird. The wing is inverted during the upstroke to form a second ring. From Pennycuick (1988).*

at the end of the downstroke, and snatch it out of the vortex ring that has just been created. The complicated 'flick' action of the wing on the upstroke was described by Brown (1948), and according to Spedding, Rayner, and Pennycuick (1984) creates no significant amount of force. Hummingbirds differ from other birds in having hardly any freedom of movement in their wing joints. At rest, the wings project out behind the bird, because they cannot be folded. In flight, the downstroke creates a vortex ring as in other birds, but the wing remains extended on the upstroke, rotates into an inverted position, and develops circulation in the opposite sense from that in the downstroke. A second vortex ring is created at the end of the upstroke, so that two rings are created in each wingbeat cycle, rather than one as in other birds (Fig. 5.15). This 'double-ring' type of wake also occurs in certain insects (Ellington 1984). Both Rayner (1979) and Ellington (1978) showed that less induced power is required to support a given weight with a double-ring wake than with a single-ring wake. Hummingbirds would appear to be specifically adapted for economical hovering, and they are the only birds known to be able to hover aerobically for prolonged periods without incurring an oxygen debt.

Wakes in cruising flight

A bird in fast flapping flight does not have to unload its wing completely during the upstroke, but it does have to generate more lift during the downstroke than during the upstroke. The lift, it should be remembered, is not necessarily directed upwards. It is defined as that component of force acting at right angles to the incident airflow. Therefore, the lift is inclined forwards during the downstroke, and backwards during the upstroke (Fig.

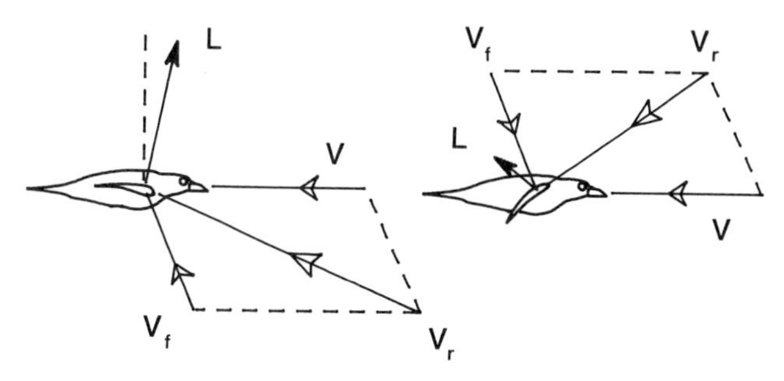

Fig. 5.16. *The lift* (L) *has to be greater in the downstroke of level flapping flight than in the upstroke, so as to develop a forward component of force, to balance the drag of the body and wings. The wing 'sees' a component of velocity* V_f *due to flapping, which comes from below in the downstroke, and from above in the upstroke. This is combined with the forward speed* (V) *to produce the resultant velocity* V_r. *Of course, the magnitude of* V_f *increases with distance from the shoulder joint, and the direction of* V_r *changes accordingly.*

5.16). Its magnitude has to be greater during the downstroke, so as to provide a forward component of force (thrust) to balance the drag of the body and the profile drag of the wings. Looking back at Fig. 5.13, there are two ways in which the bird can reduce the lift on the upstroke, by reducing the circulation or by shortening the wing span. Hummingbirds cannot shorten their wings, so they are obliged to reduce the circulation of the bound vortex during the upstroke, and therefore that of the trailing vortices also. This results in a 'ladder wake', consisting of a series of vortex rings, superimposed on the continuous, undulating wing-tip vortices (Fig. 5.17, left). The change in circulation of the trailing vortices, at the transition from upstroke to downstroke and back again, must be equal to the circulation of the vortices that form the 'rungs' of the ladder. Energy is required to create each 'rung', in proportion to its circulation and to its length, which is a little less than the wing span (Rule 3).

Quite a different type of wake was observed by Spedding (1987) in an experiment with a kestrel flying at a moderate speed. In this case the circulation of the wing-tip vortices was the same on the upstroke as on the downstroke, and the required difference in lift was obtained by shortening the wing during the upstroke (Fig. 5.17, right). This type of wake is called a 'constant-circulation wake' by Rayner (1986), or less formally, a 'concertina wake'. No energy is required to create transverse vortices with this type of wake, since none are formed, but on the other hand the energy required for the trailing vortices is a little greater than for a fixed-wing wake, because the wing span is reduced for part of the cycle. It is a reasonable supposition that the concertina type of wake is characteristic of

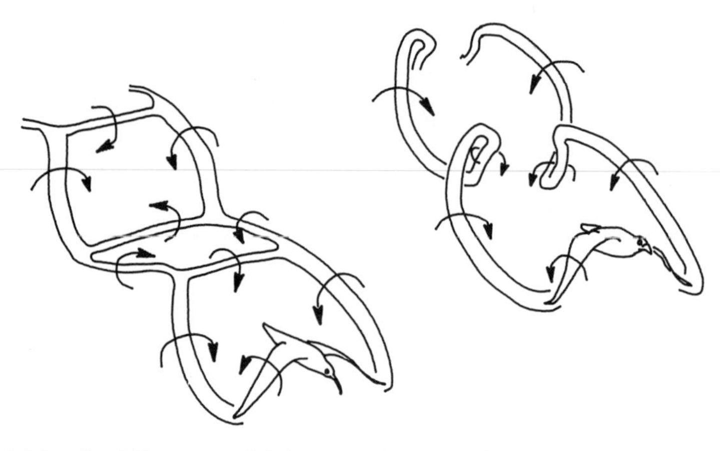

Fig. 5.17. *The difference in lift between the upstroke and the downstroke can be obtained by changing the circulation (left), in which case a transverse vortex has to be created at each change in circulation, so producing a 'ladder wake'. An alternative method is to change the span (right), keeping the circulation constant, which results in a 'concertina wake', with no transverse vortices. From Pennycuick (1988).*

birds (other than hummingbirds) in cruising flight, and is responsible for their excellent performance in long-distance migration. If this interpretation is correct, the folding type of wing may be seen as adapted for economical cruising (with a concertina wake), but with the consequence that it has to generate a single-ring wake in slow flight and hovering, which is energetically expensive. The hummingbird type of wing can generate the more efficient double-ring wake in hovering and slow flight, but cannot generate a concertina wake in cruising flight, because its span is not adjustable. Thus hummingbirds sacrifice migration performance in exchange for efficient hovering.

'Gaits' and power consumption

In a normal bird the wing action is very different depending whether it is generating a single-ring or a concertina wake. Rayner (1986) has introduced the notion that the two types of wing action, with their associated wakes, should be considered different 'gaits'. At some point a pigeon slowing down for a landing has to change over from the concertina gait to the single-ring gait, in much the same way as a horse shifts from a canter to a trot, and then a walk as it slows down. There is indirect evidence (Chapter 8) that there is an increase in the power required to fly, associated with the change of wing action, and that the shift occurs in the region of the minimum power speed. This is one reason why Programs 1 and 1A start calculating the power at a speed slightly below the minimum power speed, but do not provide power estimates for speeds lower than this. The assumption is that the induced power calculation in Chapter 3 is realistic so long as the bird is cruising fast enough to be using a concertina wake, but not if it slows down and starts generating a single-ring wake. *The methods of Chapter 3 should not be relied on for estimating induced power at speeds below V_{mp}.* The methods given by Rayner (1979) allow the induced power to be calculated for a single-ring wake, although there are some discrepancies, noted by Spedding, Rayner, and Pennycuick (1984), and Spedding (1986), which have yet to be resolved.

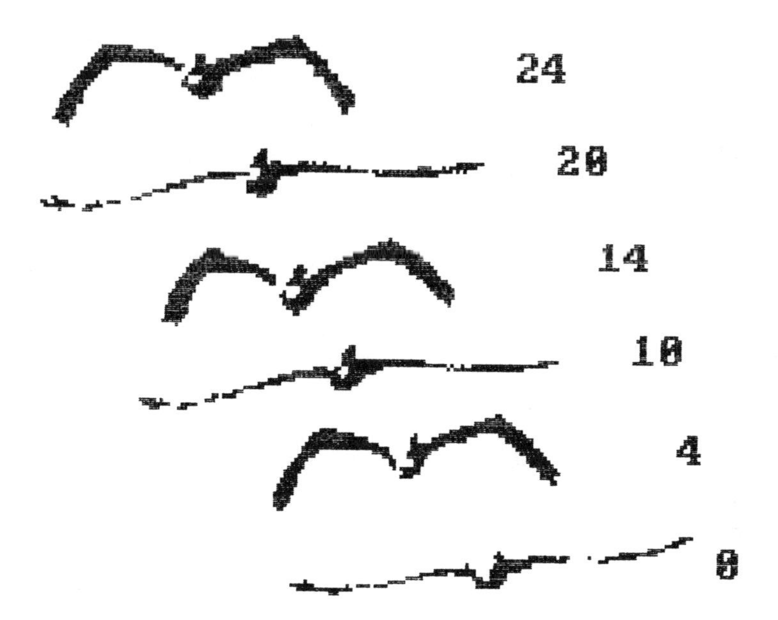

Fig. 5.18. *Video images from one second in the life of a brown pelican, frames as numbered in increments of 1/30 s. The concertina action, characteristic of cruising flight, is not easy to see with the naked eye, but is conspicuous when the action is stopped.*

6. Gliding performance calculations

Introduction

This chapter outlines a basis for calculating gliding flight performance, in a manner parallel to the flapping flight calculations of Chapter 3. The performance calculations described in this chapter are carried out by Program 2, of which a listing can be found in Appendix 1. The reader who intends to enter Program 2 from the listing is advised not to try to create it by editing Program 1 or 1A. Although the general structure of all three programs is similar, and some sections are shared by all of them, Program 2 goes off on a somewhat different track from the others at quite an early stage. It is best to make a fresh start, and enter Program 2 direct from the listing.

Gliding equilibrium

A gliding bird is one that holds its wings out steadily but does not flap them, so that it produces lift but no thrust. Gliding does not necessarily imply soaring (below). A reasonably efficient gliding wing produces a lift force (defined as that component of the aerodynamic force that acts perpendicular to the flight path) and a smaller drag force (in line with the flight path). The body also contributes drag, as in powered flight. The net force acting on a bird gliding at a steady speed must be zero (Fig. 6.1), but unlike the situation in powered flight, there is no thrust to balance the drag which (by definition) acts parallel to the incident airflow. Therefore steady gliding flight cannot be horizontal. The weight has to be balanced by the resultant of lift and drag, rather than by the lift alone. The flight path is inclined downwards at an angle α, where

$$\tan \alpha = D/L, \qquad (6.1)$$

D being the total drag and L the lift. The ratio of $L{:}D$ is the *glide ratio* or *lift to drag ratio*.

If the bird glides at a steady speed V, a component of its speed (V_z) is directed downwards. This is the *sinking speed*. In Fig. 6.1, the bird is

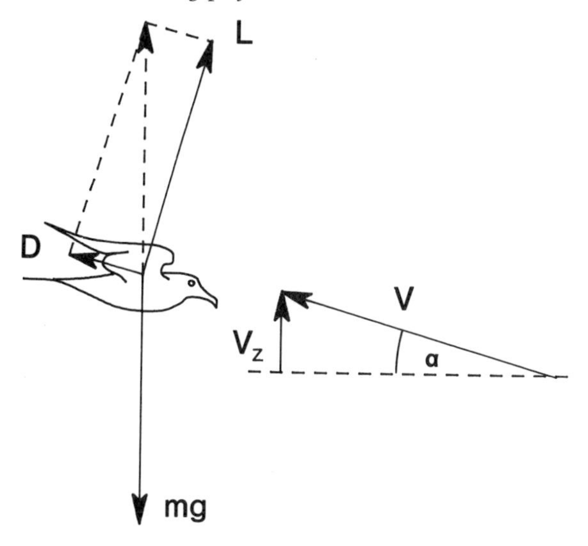

Fig. 6.1. *The weight* (mg) *of a gliding bird has to be balanced by the resultant of its lift* (L) *and drag* (D). *The flight path therefore cannot be horizontal, but is inclined downwards at an angle* α. V_z *is the vertical component of speed, or 'sinking speed'*.

considered to be stationary in a wind tunnel, with V shown as the relative wind. In this case V_z appears as an upward component of the relative wind, obtained by tilting the wind tunnel.

Minimum speed

In a steady glide, the weight is supported by the resultant of the lift and the drag, which may be called the *resultant aerodynamic force* (J). This can be represented in terms of a non-dimensional coefficient (C_J), thus:

$$C_J = 2J/\rho SV^2, \tag{6.2}$$

where S is the wing area. If the glide is not too steep, C_J is nearly the same as the lift coefficient (C_L), discussed in Chapter 5. As noted in Chapter 5, the lift coefficient is closely related to the angle through which the air is deflected, as it passes over the wing. This angle can never exceed 90°, and only comes anywhere near this value in certain aircraft that are specially designed to land and take off at very low speeds. There is some practical maximum value for the angle through which the air can be deflected, and this corresponds to a maximum value for the lift coefficient. Higher values of C_J are possible if the wing generates vast amounts of drag, so that the bird comes down in a very slow, steep descent. The mechanics of this type of flight have been considered by Tucker (1987), but are outside the main area of interest here. For present purposes, it is a sufficiently close approximation to assume that the weight is balanced by the lift (rather than the resultant force), in which case:

$$C_L = 2mg/\rho SV^2. \tag{6.3}$$

If C_L has some maximum value (C_{Lmax}), then we can turn equation (6.3) round to give the minimum gliding speed:

$$V_{min} = \sqrt{(2mg/\rho SC_{Lmax})}. \tag{6.4}$$

Tucker (1988) reported lift coefficients up to 2.2 from field observations of African white-backed vultures, but there was considerable margin for error in the measurements. Wind tunnel tests of gliding pigeons (Pennycuick 1968) and falcons (Tucker and Parrott 1970), both yielded lift coefficients up to about 1.6 for gliding flight at modest angles. At lower speeds the birds preferred to change over to flapping flight.

Fixed span glide polar

In stationary air, the bird loses potential energy at a rate equal to its weight multiplied by the sinking speed. This 'force-times-speed' can be considered the power that has to be supplied by gravity, and it has to balance the power needed to overcome the drag, that is,

$$mgV_z = DV, \tag{6.5}$$

where m is the body mass and g is the acceleration due to gravity. The drag is itself a function of speed. If we want to plot a curve of sinking speed versus forward speed, we first have to express the drag as a function of speed, and then get the sinking speed by rearranging equation (6.5) thus:

$$V_z = DV/mg. \tag{6.6}$$

A curve of sinking speed versus forward speed is called a *glide polar*, and is the basis of gliding performance calculations. The sinking speed is directly equivalent to the power calculated in Chapter 3—actually it is the power divided by the weight, as equation (6.6) shows. A diagrammatic glide polar is sketched in Fig. 6.2. It shows the two characteristic speeds, the *minimum sink speed* (V_{ms}), corresponding to the minimum power speed of the flapping flight curve, and the *best glide speed* (V_{bg}), corresponding to the maximum range speed. The *minimum speed* (V_{min}) is determined by the maximum lift coefficient (above).

The usual way to calculate a glide polar is to add together three separate components of drag; the parasite drag, the induced drag, and the profile drag. These are forces, and they have to be multiplied by the forward speed to get power, which is proportional to the sinking speed. The three

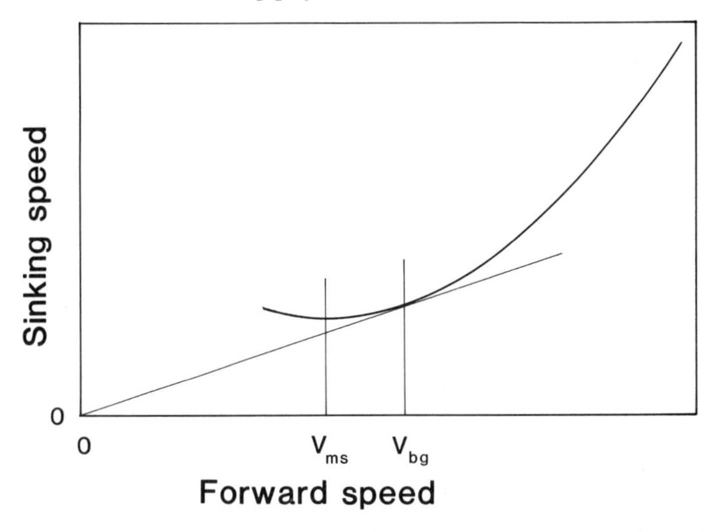

Fig. 6.2. *The glide polar is a plot of sinking speed versus forward speed. Glider pilots often plot polars with the y-axis going downwards (because sink is down). When plotted this way up, the glide polar looks very much like the power curve of Fig. 1.1, and is closely analogous to it.* V_{ms} *is the speed for minimum sinking speed, and* V_{bg} *is that for best gliding angle.*

components correspond closely to the parasite power, induced power, and profile power of Chapter 3, and the method of calculating them is basically the same. The basis of the calculation performed by Program 2 is as follows.

Parasite drag

The drag of the body is found in the same way as in Chapter 3. First, the body frontal area is found from equation (3.5), then the drag coefficient is estimated on the basis of a rough estimate of the Reynolds number in cruising flight, as in Fig. 5.7. If the user is not content with the calculated values, they can be changed from the main menu of Program 2. As in Chapter 3, the parasite drag is

$$D_{par} = \tfrac{1}{2}\rho S_b C_{Db} V^2, \qquad (6.7)$$

where the body's frontal area is S_b and its drag coefficient is C_{Db}.

Induced drag

At this stage a fixed value (b) is taken for the wing span, although later this will be modified to allow for a gliding bird's habit of reducing its wing span at high speeds. Whether b is fixed or not, classical aerodynamics indicates that the induced drag (D_{ind}) associated with a given value of b is:

$$D_{ind} = 2km^2g^2/\pi\rho V^2 b^2, \qquad (6.8)$$

where k is a dimensionless 'induced drag factor' (greater than 1), and ρ is the air density. This result is derived in classical textbooks such as that of von Mises (1945), and is based on the assumption that the wing creates a vortex wake, like that of Fig. 5.13, in the most efficient possible way. The reader may like to multiply the induced drag from equation (6.8) by the speed, to turn it into induced power, and compare the result with the induced power calculated for flapping flight in equation (3.10). The latter contains the disc area (S_d), which has to be expressed in terms of the span (from equation (3.6)). The implications of this comparison have been mentioned in Chapter 3.

The default value for k in Program 2 is taken to be 1.1, rather than 1.2 for the corresponding variable in the flapping flight case. This reflects a general impression that the vortex wake of a gliding bird more closely approximates the ideal form than that of a flapping bird. However, the 'concertina' type of flapping wake may be more efficient than would be suggested by the choice of default values in Programs 1 and 1A. Some recent field data suggest that k should be reduced to 1.1 for that case also (Chapter 8). The user can, of course, change the value of k from the main menu in any of the programs.

Profile drag

Profile drag is well documented for gliders, but unfortunately the scale effects discussed in Chapter 5 make the transfer of the results to birds difficult and uncertain. The basis of calculating profile drag (D_{pro}) is more direct than the profile power calculation of Chapter 3. It is found by using a dimensionless *profile drag coefficient* (C_{Dpro}), thus:

$$D_{pro} = \tfrac{1}{2}\rho S C_{Dpro} V^2, \qquad (6.9)$$

where S is the wing area. As noted in Chapter 5, it is known that many (perhaps all) gliding birds habitually fly around with detached boundary layers in slow flight, and a detached boundary layer over a wing results in a high profile drag coefficient. For the practical calculations performed by Program 2, it is assumed that C_{Dpro} is independent of speed, although it is recognized that this is not a good assumption at low speeds. However, the purpose of Program 2 is to predict gliding performance when the bird is gliding at medium and high speeds. As in the case of flapping flight, an approximation is used which has a chance of being realistic, provided it is not used outside the range of speeds for which the program is intended. At low speeds (high lift coefficients), the profile drag coefficient is likely to increase sharply, mainly on account of the tendency of the boundary layer to separate. Thus, if the program gives realistic results at medium and high speeds, it must be expected that the profile drag (and sinking speed) will be underestimated near the minimum speed. The program starts calculating the glide polar at a speed corresponding to $C_L = 1.6$, and is likely to underestimate the sinking speed at the low-speed end of the curve. The calculation as it stands in Program 2 should not be extrapolated to lower speeds.

Total drag

The total drag (D) is found by adding together the three components calculated from equations (6.7), (6.8), and (6.9), thus:

$$D = D_{par} + D_{ind} + D_{pro}. \tag{6.10}$$

To calculate the glide polar, the drag is calculated for each value of the speed from equation (6.10), and then used to find the sinking speed from equation (6.6).

Gliding performance with variable wing span

When gliding at high speeds, birds flex their elbow and wrist joints in a manner that reduces both the span and the wing area (Figs. 6.3 and 6.4). The reduction of area is due to each flight feather increasing the amount of overlap with its neighbours. The chord of the wing is not greatly changed

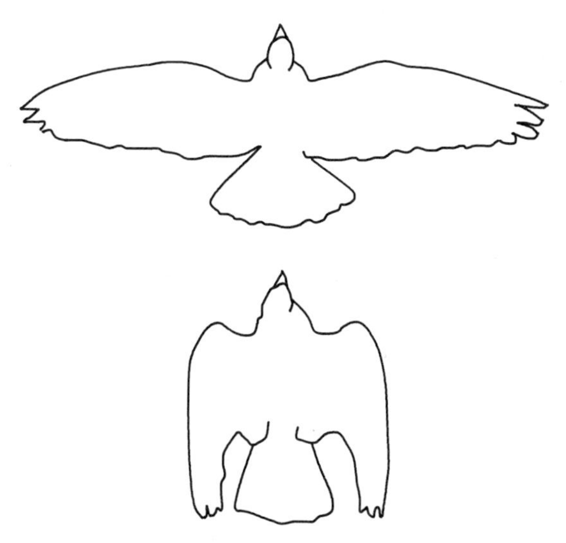

Fig. 6.3. *Silhouettes of a pigeon gliding in a tilted wind tunnel, at a low speed (top) and a high speed (bottom). As speed increases, the pigeon reduces both wing span and area, by flexing the elbow and wrist joints. After Pennycuick (1968a).*

Fig. 6.4. *A black-browed albatross with its wings flexed for fast gliding, as in Fig. 6.3. South Georgia.*

by this. The wing span and area, b and S as defined in Chapter 2, are the values measured with the wing fully spread. If the wing span is now reduced by a factor β, this results in the area being reduced by some factor ϵ, as shown in Fig. 6.5. If this graph is considered to be a straight line (which is a sufficiently precise assumption for present purposes), then its equation is

$$\epsilon = \{1-\delta(1-\beta)\}. \tag{6.11}$$

Only the *planform slope* (δ) needs to be specified in order to define the line, since it has to pass though the point $\epsilon=1$ when $\beta=1$. In Program 2, the planform slope is taken to be 1, meaning that the mean chord of the wing remains constant as the span is reduced. It is difficult to measure δ from a specimen, as one cannot be sure exactly how the bird would flex the joints in flight. However, measurements on a dead Harris hawk, with the wings set in a posture resembling that seen in hawks when gliding fast, gave $\delta=0.93$, while Tucker (1988) gives figures for the African white-backed vulture which yield $\delta=1.06$. The latter implies that the mean chord actually decreases when the span is reduced, which is difficult to imagine anatomically. The reader can, of course, install any value of δ from the main menu when running Program 2.

Reducing the wing span increases induced drag (equation (6.8)). Since the induced drag is inversely proportional to the speed-squared, the drag

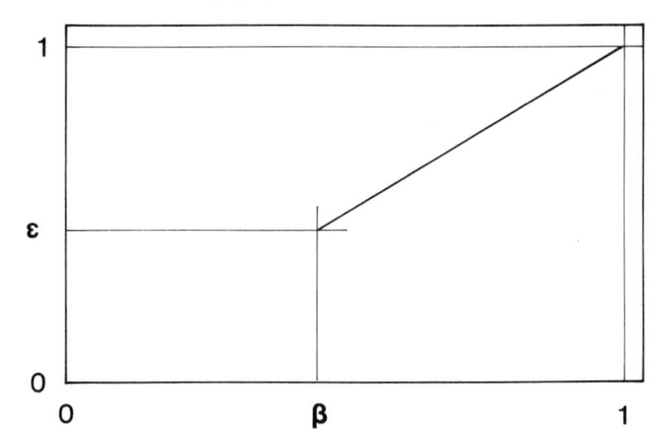

Fig. 6.5. *Graph to describe the change of shape in Figs. 6.3 and 6.4. ϵ is the ratio of the actual wing area to the value with the wing fully spread, and β is the ratio of the actual wing span to its maximum value. Since the line has to pass through $\epsilon = 1$ when $\beta = 1$, only one number (the slope) is needed to define the line. This is the* planform slope.

penalty, caused by reducing the span, is large at low speeds, and decreases progressively at higher speeds. The associated reduction of wing area results in a reduction of profile drag (equation (6.9)). In this case, the effect is greater at high speeds, because profile drag varies directly with the speed-squared. At low speeds, the effect of reducing the wing span is to cause a large increase in induced drag, and a small reduction of profile drag. At high speeds, the effect may initially be to reduce the profile drag by an amount exceeding the increase in induced drag. Eventually, continued reduction of the span will lead to an increase in induced drag that is larger than the reduction of profile drag. This defines an optimum wing span, less than the maximum, at which the sum of the induced and profile drags is at a minimum.

A glide polar can be calculated for any value of the span factor (β) by substituting βb for b in equation (6.8), and ϵS for S in equation (6.9), where ϵ is calculated from b via equation (6.11). The glide polar then becomes:

$$V_z = (2kmg/\pi\rho\beta^2 b^2 V) + \{\rho V^3 (C_{\text{Dpro}}\epsilon S + S_b C_{\text{Db}})/2mg\}. \qquad (6.12)$$

Notice that the weight (mg) appears in the *denominator* of the second term of equation (6.12), which combines the contributions due to parasite and profile drag. It may seem odd that increasing the weight should reduce the rate of sink, but it does so at high speeds. Racing glider pilots habitually carry large quantities of water in ballast tanks in the wings of their machines. The added weight does indeed reduce the rate of sink, when gliding at high speeds.

Figure 6.6 shows at left a polar calculated for the American turkey

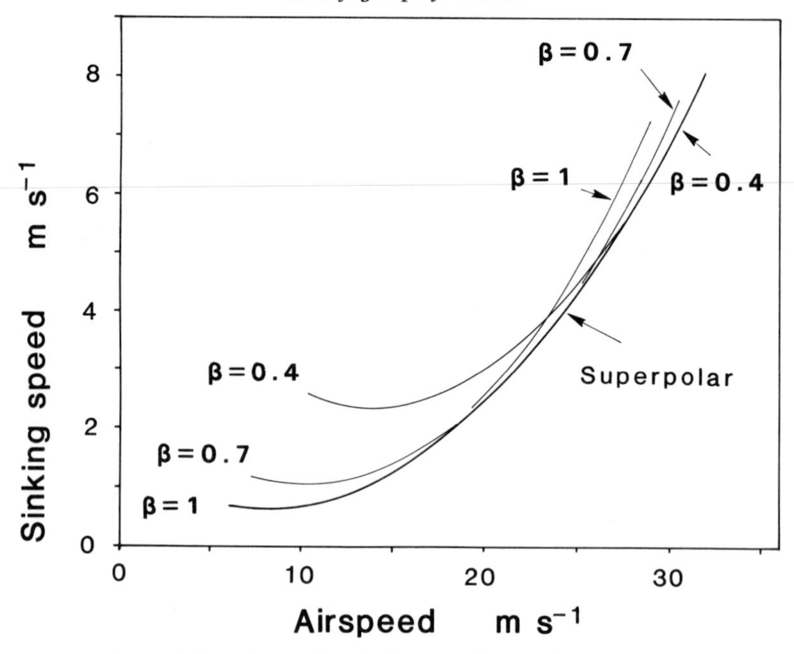

Fig. 6.6. *Three glide polars plotted for a turkey vulture, with its wing span reduced by various amounts. At high speeds, the shorter wing span works better. The* superpolar *is the envelope of this family of fixed-wing polars, choosing the value of β that gives the least sink at each speed.*

vulture with the wings fully spread ($\beta=1$), and further to the right, polars with $\beta=0.7$, and $\beta=0.4$. At low speeds, full span gives the lowest sinking speed, while the two curves with reduced span give more sink, because of the higher induced drag. At speeds above about 18 m s^{-1}, the 70 per cent span curve is better than the full-span curve, because the induced drag is less at this speed, and the reduction of span produces a larger reduction of profile drag. At still higher speeds, above about 25 m s^{-1}, the 40 per cent span curve is the best.

Glide superpolar

Tucker and Parrott (1970) measured the glide polar of a falcon flying in a tilting wind tunnel, and noted that the bird progressively reduced its span as the wind speed was increased, so as to trade off induced against profile drag, and to obtain the minimum rate of sink at each speed. The bird's glide polar effectively became the envelope of the separate fixed-span polars shown in Fig. 6.6. This is the 'superpolar', in which the bird selects the optimum wing span to obtain minimum drag at each speed. At the low-speed end, the superpolar coincides with the full-span polar, because the bird cannot stretch its wings out to a greater span than b. Tucker (1987) has considered the theory of the superpolar, along with that of high-drag configurations which do not concern us here. For present purposes it is sufficient to note that the value of β which gives the minimum rate of sink from equation (6.12) is

$$\beta = \{(8km^2g^2)/(\delta\pi\rho^2b^2SC_{\mathrm{Dpro}}V^4)\}^{1/3}. \quad (6.13)$$

At low speeds, equation (6.13) will indicate that the optimum value of β is greater than 1, which is not permitted for anatomical reasons. Program 2 first calculates β from equation (6.13). If the result is greater than 1, then both β and ϵ are set to 1 (full span and area). Otherwise, ϵ is calculated from β (equation (6.11)), and then the sinking speed is found from equation (6.12). Program 2 does this calculation. It prints out the superpolar in the form of a table of sinking speed versus forward speed, before proceeding to a cross-country speed calculation (below).

Soaring

Soaring is any form of flight behaviour whereby a bird or pilot extracts energy from movements of the atmosphere. It is possible for a bird to soar and flap its wings at the same time, using atmospheric energy to supplement its own muscular efforts. However, birds that are specialized for soaring normally do so in gliding flight, and this has much in common with the soaring flight of sailplanes. The glide polar is the basis of soaring flight performance calculations. For birds, the superpolar of equation (6.12) can be used.

Slope soaring

The simplest form of soaring is *slope soaring*, where the source of energy is rising air, caused by the wind being deflected upwards by a slope. For glider pilots, a 'slope' means a ridge or a mountainside, but birds such as gulls and turkey vultures can slope-soar along buildings and trees, while petrels and albatrosses use waves on the surface of the sea as soarable slopes. Slope soaring is used as a method of migration in some parts of the world, notably by raptors that follow the Appalachian ridges in eastern North America (Kerlinger 1989). The 'lift', which in this context means the upward component of the wind speed, is usually strongest at or near ridge-top height. Consequently, streams of migrants may be seen flying along the crests of the ridges, much to the satisfaction of hawk-watchers on the ground.

For slope soaring to be possible, the strength of the lift must at least

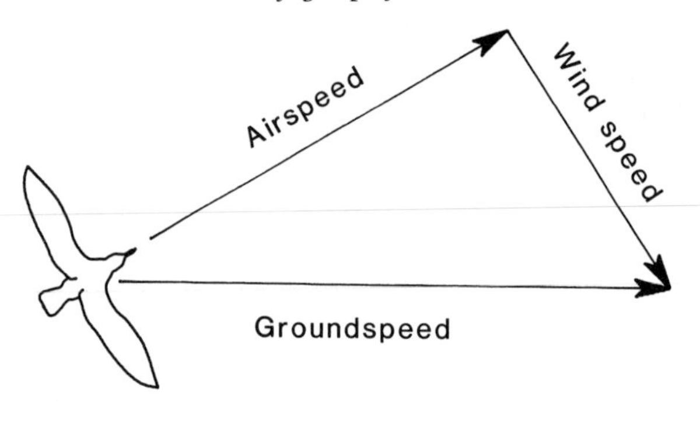

Fig. 6.7. *When a migrating bird slope-soars along a ridge, the groundspeed observed by stop-watch methods is the vector sum of the airspeed and windspeed. The airspeed is needed to get the sink from the superpolar.*

equal the bird's minimum sinking speed (Fig. 6.6). If the lift is stronger, the bird can fly, without losing height, at a speed that makes its sinking speed equal to the lift strength. If the user has an estimate for the lift strength to be found on a particular ridge, then the airspeed at which the bird can fly can be read off the superpolar, as printed out by Program 2. This speed is not identical to the groundspeed at which the bird will travel along the ridge. For that, the triangle of velocities has to be solved as in Fig. 6.7. Gadgets for solving problems of this type are sold in pilot shops.

Thermal soaring

Thermals are vortex structures in the atmosphere that contain rising air. Thermals may occur over flat or mountainous country, on still or windy days. They are often triggered by solar heating of the ground, and may extend to various heights. Thermals can never be relied upon to occur at all, but when they occur over land, they usually build up as the ground warms up during the day, and die down when it cools at night. Thermals also occur over the sea, in the trade-wind zones, but here the mechanism is different. Heating at the base of the atmosphere is due to the air being carried towards the equator by the trade winds, so bringing it over progressively warmer water. Diurnal variation of the water temperature is very slight, and so trade-wind thermals continue both by day and by night, and the associated cumulus clouds look much the same by moonlight as they do by day.

Many raptors and vultures soar in thermals over land as a means of patrolling in search of food, and frigatebirds do much the same thing over the sea. These and other medium-sized or large birds, such as storks, cranes, and pelicans, also use thermals for travelling across country, either on migration or on foraging flights when nesting. Thermals are used like stepping stones for cross-country flight. Having found a thermal, the bird climbs in it until it can get no higher, then glides off straight, losing height, and hoping to find another thermal before being obliged to land, or resort to flapping flight. The theory of cross-country flight in thermals has been well studied, since gliders use essentially the same method. The process is usually represented as a series of cycles, each cycle consisting of a climbing phase followed by a gliding phase. Real cross-country flights do not invariably conform to this pattern, but it is realistic enough to give a good basis for calculating cross-country speed from the glide polar.

Climbing in thermals

To climb in a thermal, the bird has to fly around in circles so as to stay within a limited area of lift, and it also has to keep its sinking speed as low as possible. The lift in a thermal is strongest in a 'core', which may be quite narrow, and weaker round about. In some kinds of thermal, the core is surrounded by an area of negative lift ('sink'). If the thermal is narrow, the bird has to keep its radius of turn small enough to stay in the lift. The bird does this by gliding with its wings inclined at an *angle of bank* (ϕ) to the horizontal, in which case the radius of turn (r) is

$$r = 2mg/\rho gSC_L \sin \phi. \tag{6.14}$$

Evidently the radius of turn at any particular lift coefficient is directly proportional to the wing loading (mg/S), other things being equal. Three different (and morphologically dissimilar) species observed by Pennycuick (1983) in Panama did indeed circle at average radii that were proportional to their wing loadings, varying from 12 m for a frigatebird to 18 m for a brown pelican. The lift coefficients and angles of bank were nearly the same for all (about 1.4 and 24°). The theory of circling flight (outlined by Pennycuick 1975) considers how the sinking speed increases as the radius of turn is reduced by steepening the angle of bank, and theoretically becomes infinite when the wings are vertically banked. The radius does not approach zero at very steep angles of bank, but asymptotically approaches a limiting value (r_{lim}), where

$$r_{lim} = 2mg/\rho gSC_L. \tag{6.15}$$

This limiting radius is proportional to the wing loading. Smaller radii are inaccessible, even under theoretically extreme conditions of flight. Because of simple scale effects, gliders have higher wing loadings than birds, and are obliged to fly in larger circles. It often happens that birds can circle easily in a thermal that is below the limiting radius for a glider. In such narrow thermals, birds can soar, whereas gliders cannot. In slightly wider thermals, birds can still outclimb a glider by circling in a narrow core, while the glider is forced to use weaker lift around the periphery. However, once the thermals grow beyond a certain size and strength, differences in gliding performance cease to have much effect on the rate of climb. It is a surprising feature of thermal soaring, often noticed by glider pilots, that the practical rate of climb in large, strong thermals does not seem to vary a great deal for gliders old or new, birds large or small, or even such random objects as plastic bags and displaced items of laundry.

Circling performance can be calculated in the form of a curve relating sinking speed to circling radius (Pennycuick 1975). However, this involves estimating the drag at very high lift coefficients, which, as noted above, is a procedure fraught with uncertainty. The performance estimates, once obtained, refer to sinking speed in still air. To translate them into rate of climb, additional information is needed about the distribution of vertical velocity across the core of the thermal in which the bird is endeavouring to soar. Naturally, one cannot consider individual thermals. Climbing performance can only be calculated in hypothetical thermals with specified characteristics, whose resemblance to the real thing is bound to be approximate at best. Program 2 does not calculate climbing performance, except to print out a representative circling radius for 24° of bank and $C_L = 1.4$. Birds for which this radius is small may be regarded as well adapted to exploit narrow thermals. The actual climbing performance of different species depends very strongly on the characteristics of the thermals, which are highly variable from place to place and from minute to minute.

Cross-country flight

Program 2 does not attempt to calculate rate of climb in thermals. Instead, the achieved rate of climb in thermals is treated as an independent variable, representing the 'weather'. The superpolar is used to calculate the speed at which the bird travels across country, given a particular rate of climb in thermals. This calculation is based on a classical gliding theorem, illustrated in Fig. 6.8. The superpolar is plotted as in Fig. 6.6, and the y-axis is extended downwards, where it represents rate of climb in thermals. If the bird climbs at the rate V_c, marked on the extended y-axis of Fig. 6.8, and elects to glide at the *inter-thermal speed* (V_{it}), then the average speed at which it makes progress across country is V_{xc}, found by noting where the straight line, drawn as shown, crosses the x-axis. Given this result, it follows that

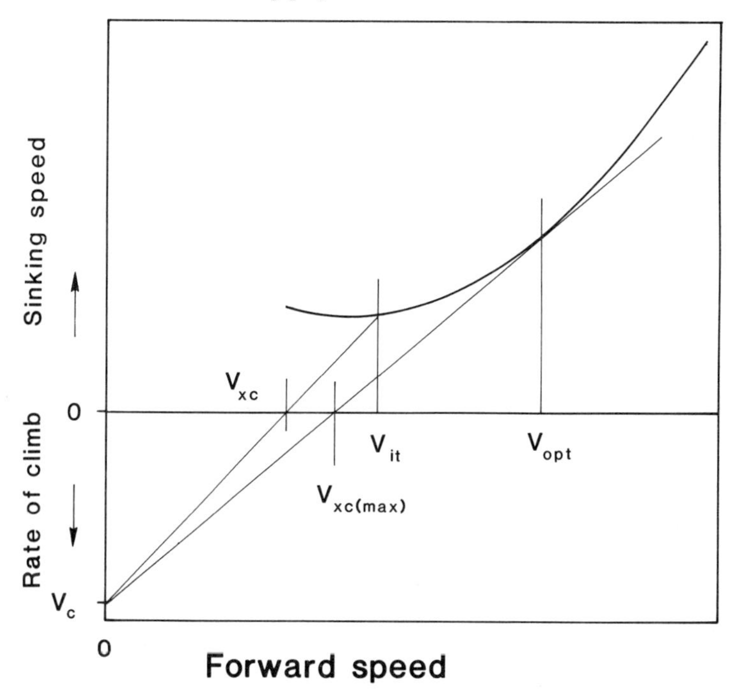

Fig. 6.8. *Construction for finding a bird's cross-country speed* (V_{xc}), *if it achieves a rate of climb* V_c *in thermals, and elects to fly at some arbitrary inter-thermal speed* (V_{it}) *when gliding from one thermal to the next. For the given rate of climb, the maximum cross-country speed is obtained by flying between thermals at a speed* (V_{opt}) *found by drawing a tangent to the glide polar as shown.*

the optimum value of the inter-thermal speed (V_{opt}), can be found by drawing a tangent to the superpolar, from the point on the extended y-axis representing the rate of climb V_c.

Program 2 first selects a value for the rate of climb, starting with 0.5 m s^{-1}. It then works its way along the superpolar, increasing the speed in steps of 0.1 m s^{-1}. After each step, it projects a tangent back to the extended y-axis, and checks to see whether the tangent intercepts the axis at a value of V_c equal to or above the currently selected value. As soon as this occurs, V_{opt} is set to the current value of the speed, and V_{xc} is found by the construction of Fig. 6.8. The rate of climb is then incremented by 0.5 m s^{-1}, and the computation of the superpolar continues until V_{opt} is found for this new rate of climb. This goes on at the same time that the superpolar is being calculated and tabulated, but the results of the cross-country speed calculation are presented as a separate table.

Options in choosing inter-thermal speed

Program 2 also prints out a second set of cross-country speeds, calculated on the assumption that the bird elects to fly between thermals at the best-glide speed (V_{bg}, as marked in Fig. 6.2). These cross-country speeds are, of course, lower than those obtained by flying at V_{opt}. It might appear that there would be no reason to fly at any speed other than V_{opt}. Although glider pilots carry a cockpit instrument that calculates V_{opt} for the prevailing conditions, they often elect to fly at a slower speed between thermals, even when trying to maximize cross-country speed. The reason is that the optimum speed is often quite high if thermals are strong, and flying at a high speed results in descending at a steep angle. This can be seen from the last column of the superpolar table produced by Program 2, which prints out the glide ratio at each speed.

To take an example from the specimen output for the white-backed vulture in Table A1.9, suppose that the vulture arrives at the top of a thermal, having achieved a rate of climb of 4 m s^{-1}. The optimum inter-thermal speed is listed as 26.6 m s^{-1}, and if the vulture sets off for the next thermal at this speed, it should realize a cross-country speed of 14.6 m s^{-1}. The fourth column shows that if it flies at the best-glide speed (13.5 m s^{-1}), regardless of the rate of climb, the cross-country speed would be only 11.0 m s^{-1}. However, the superpolar table shows that the glide ratio is 14.6 at 13.5 m s^{-1}, and only about 8 at 26.6 m s^{-1}. When the vulture leaves its thermal, it can only afford to lose a certain amount of height before it needs another thermal, and to maintain its cross-country speed, the next thermal must be as strong as the last one. By electing to fly at V_{bg} rather than at V_{opt}, it can glide 1.8 times as far before it needs another thermal, and this increases the chance of finding another strong thermal for the next climb. If the bird comes down too steeply, as a result of setting off from the top of the last thermal at an imprudently high speed, it may be forced to accept a weak thermal for the next climb. Suppose that it has to accept a 2 m s^{-1} thermal, then the table shows that the best cross-country speed it can now achieve is only 10.6 m s^{-1}, even if it flies at the new V_{opt} between thermals. Thus the prudent bird or pilot only flies as fast between thermals as theory recommends, if thermals are strong, frequent, and clearly marked by cumulus clouds, and if plenty of height is available above the ground. Otherwise, it is better to choose a slower inter-thermal speed. There is never any advantage in flying slower than V_{bg} or faster than V_{opt}. Since this theory depends on calculating the superpolar from V_{bg} upwards, the approximation that the profile drag coefficient is constant, as noted above, can be accepted.

Fig. 6.9. *Common cranes migrating over southern Sweden, gliding between thermals in characteristic 'vee' formation.*

Powered inter-thermal glides

Migrating cranes vary the simple method of cross-country soaring, described above, by using variable amounts of power during the inter-thermal glides. If conditions are poor, a flock of cranes will extend the glide by flapping intermittently and at low amplitude. The lower they get, the more power they use, until eventually, if no thermals are found, they find themselves flapping along in level powered flight over the treetops. Pennycuick, Alerstam, and Larsson (1979) devised an extension of the traditional theory of cross-country flight to apply to this strategy, showing that there is a continuous tradeoff between the amount of energy consumed on a flight, and the time taken to complete it. Steady flapping is fastest, but energetically expensive, whereas pure soaring is most economical, but slow. This theory is not covered in Program 2, although it could be incorporated.

Fuel consumption in soaring

In addition to its basal metabolism, a gliding bird requires energy for the tonic muscles that hold the wings in the horizontal position. Except in albatrosses and giant petrels, there is no mechanical lock to prevent the outstretched wings from rising up during gliding, and therefore continuous muscular activity is needed to hold the wings in position. The power required for this is considered in the next chapter, where somewhat insecure grounds are put forward for supposing that this component of power might well be proportional to the basal metabolic rate in birds of different size. A solitary physiological measurement on a gull by Baudinette and Schmidt-Nielsen (1974) yielded an estimate of about twice the basal metabolic rate for the bird's total power consumption when gliding in a wind tunnel. Program 2 calculates the power consumption on this basis, and uses the result to supply an estimate of the amount of fat consumed per unit distance flown, when soaring across country in thermals.

7. Power available from the muscles

Introduction

The calculations of Chapter 3 are concerned with calculating the power *required* to fly horizontally at a specified speed. The bird is only capable of flying at that speed, if at least the indicated amount of power is *available* from its muscles. The power output of a muscle, that is the rate at which it does mechanical work, is a purely mechanical matter. The upper limit to the power that a muscle can produce is set by the force which it can exert, the distance through which it can shorten, and the frequency at which it can contract. The chemical systems that process energy in the muscle have to be adapted to supply energy at a rate at which the contractile machinery can use it. No amount of enzyme activity can increase the power output of a muscle above the maximum set by mechanical factors.

This chapter is concerned with the basic principles of the generation of force, work, and power by muscles. Bird flight involves muscles of two broad types, those that produce the power for flapping flight, and those that produce a steady force to hold the wings level in gliding flight. A distinction between aerobic and anaerobic muscles is also needed. Beyond this, the reader is referred to George and Berger (1966) and Alexander (1968) for the many types of mechanical and histological adaptations seen in muscles adapted to different functions. White and Thorson (1975) give a good explanation of the mechanism of muscular contraction at the molecular level. The whole field was reviewed by McMahon (1984).

Readers should note that some algebraic symbols (a, b, α, β, ν) have meanings that are different in this chapter from elsewhere in the book. Both meanings are listed in Appendix 3. This should not cause confusion, so long as the duplication is noticed.

Work and power output of the myofibrils

Work done by a muscle

Figure 7.1 represents a simplified muscle. It could be large or small, and might be a single fibre or even part of a fibre. It contains a parallel array

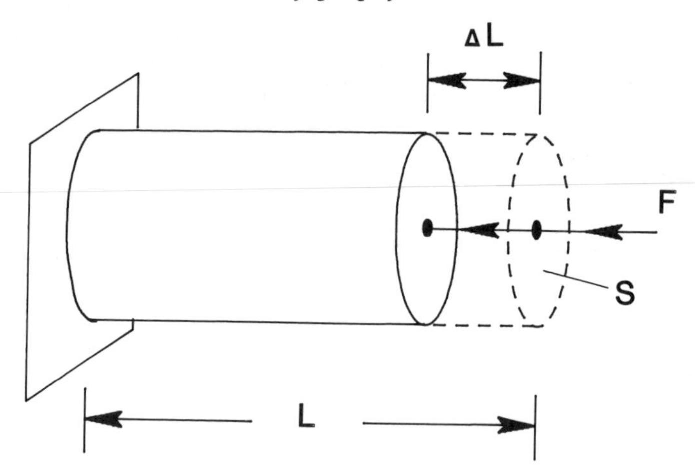

Fig. 7.1. *A muscle of extended length* L *and cross-sectional area* S *shortens through a distance* ΔL, *exerting a force* F *as it does so. From Pennycuick and Rezende* (1984).

of contractile filaments which, when activated, exert a force (F) over a cross-sectional area (S). Its initial length is L, but it can contract through a distance ΔL. When it does so, the work done (Q) is

$$Q = F\Delta L. \tag{7.1}$$

The 'raw' variables, force and distance, are appropriate for describing an individual muscle, but inconvenient for expressing more general properties that apply to all muscles, large or small, of a given type. For this purpose, it is better to replace force by stress, and distance by strain. The *stress* (σ) that a muscle exerts is the force per unit cross-sectional area. In terms of the diagram of Fig. 7.1,

$$\sigma = F/S. \tag{7.2}$$

The *active strain* (λ) is the distance shortened, expressed as a proportion of the extended length:

$$\lambda = \Delta L/L. \tag{7.3}$$

The dimensions of stress are $ML^{-1}T^{-2}$ (the same as pressure), while strain is dimensionless. If the stress and strain are multiplied together, the result is:

$$\sigma\lambda = F\Delta L/SL = Q/v. \tag{7.4}$$

The numerator in equation (7.4) is equal to the work done (from equation (7.1)), while the denominator is the extended length times the cross-sectional area, that is the volume (v). In other words, the product of the stress and the strain gives the *volume-specific work* (Q_v), or work done by unit volume of muscle tissue. The *mass-specific work* (Q_m) can be obtained from the volume-specific work by dividing it by the density of the muscle (ρ):

$$Q_m = Q_v/\rho = \sigma\lambda/\rho. \tag{7.5}$$

Regardless of the size or shape of the muscle, the stress exerted by its fibres, times the active strain, divided by the density, gives the work done per unit mass of muscle, in one contraction. The stress and the strain are readily understandable at the microscopic level. If the force is being exerted by a parallel array of myofibrils, the stress is proportional to the force exerted by each myofibril. A typical value might be 500 pN for each myosin filament in a muscle exerting an isometric stress of 300 kPa. The strain expresses the extent of overlap between the actin and myosin filaments as the muscle shortens.

Power developed by a muscle

Locomotor muscles operate in a cyclic fashion. The muscle contracts, doing work, and is then passively stretched out before it is ready to do more work in another contraction. The power output, or rate of doing work, is simply the amount of work done in each contraction, times the contraction frequency (f). The *volume-specific power* (P_v) is the product of the volume-specific work (equation (7.5)) and the contraction frequency:

$$P_v = Q_v f = \sigma\lambda f. \tag{7.6}$$

The mass-specific power can be found by dividing the volume-specific power by the muscle density:

$$P_m = P_v/\rho = \sigma\lambda f/\rho. \tag{7.7}$$

The stress, strain, and density are properties which would not be expected to vary much in muscles of different size or shape, but of the same general type, and performing similar tasks. The contraction frequency, however, is lower in large animals than in small ones, and therefore so is the mass-specific power output. The mass-specific power output in turn is closely related to the mass-specific rates at which the muscle consumes fuel and oxygen, and generates heat. A microscopic observer embedded in the flight muscles of a bird, measuring the rate at which the fibres consume energy, could very easily distinguish whether he was in a swan or a hummingbird, because *each gram* of a hummingbird's muscles consumes energy at a much higher rate than those of a swan. In aerobic muscles, this is reflected in the provision of blood vessels and mitochondria, which are much more prominent in hummingbird muscles than in those of larger birds.

Dynamic properties of muscles

The dynamic properties of locomotor muscles have to be adapted to suit the size of the animal they propel. This is ultimately a biochemical matter, but the nature of the adaptation may be understood in terms of mechanical properties. Hill (1938) first discovered the 'force–velocity relationship' that governs the mechanical capabilities of any particular muscle. This was based on measurements of the speed at which a muscle shortened in an 'isotonic contraction', that is, while pulling against a constant force, maintained by the apparatus. For any particular muscle, there is some force (F_0) against which the muscle cannot shorten at all, that is, the speed of shortening is zero. This is the *isometric force*. If more force is applied, the muscle is forcibly stretched out. At the other end of the graph, if the muscle is allowed to shorten against zero force, the speed of shortening reaches some maximum value (V_{max}). In between, the velocity is related to the force by 'Hill's equation':

$$V = b(F_0 - F)/(F + a). \quad (7.8)$$

This equation describes a hyperbolic curve, as in Fig. 7.2. The value of V when $F = 0$ is

$$V_{max} = bF_0/a \quad (7.9)$$

Equation (7.8) is not just the result of an arbitrary curve-fitting exercise. Hill identified the dimensions of the constants a and b (force and velocity respectively), and carefully interpreted their physical significance, in terms of the production of heat and work during contraction. His 1938 paper is

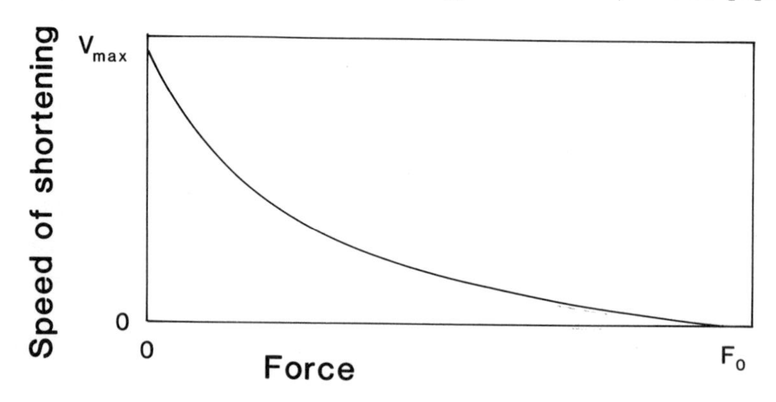

Fig. 7.2. *The relation between speed of shortening and force in an isolated muscle shortening against a constant force. The curve is described by Hill's equation (7.8).* V_{max}, *speed of shortening against zero force:* F_o, *isometric force, with the length held constant.*

a model of meticulous attention to physical dimensions, setting a standard which, regrettably, has not always been maintained during the later development of physiology.

Hill's purpose in studying the force–velocity relationship was to elucidate the functioning of muscle. The present purpose is to represent the properties of muscle in relation to its use in locomotion. For this, it is more convenient to express Hill's equation in terms of stress and strain, rather than force and speed. The original form (equation (7.8)), which relates speed to force, can be replaced by a form of the same equation in which speed is replaced by the *strain rate* (ν) (that is the rate of change of strain), and force by stress:

$$\nu = d\lambda/dt = \beta(\sigma_0 - \sigma)/(\sigma + \alpha). \tag{7.10}$$

The constants α and β are derived from Hill's constants a and b thus:

$$\alpha = a/S, \text{ and } \beta = b/L. \tag{7.11}$$

The dimensions of the new constants are different from those of the old ones. α has the dimensions of stress ($ML^{-1}T^{-2}$), while β has the same dimensions as strain rate (T^{-1}).

The graph of this form of Hill's equation looks the same as the original form (Fig. 7.3). The advantage of changing the dimensions is that whereas the original isometric *force* is large for a large muscle, and small for a small one, the isometric *stress* (σ_0) is much the same for all muscles of the same general type. The x-axis of Fig. 7.2 can only be numbered if the diagram refers to a particular individual muscle, but the stress values on the x-axis of Fig. 7.3 are as realistic for any vertebrate locomotor muscle as for any other. We can put the isometric stress at 300 kPa if the cross-sectional area

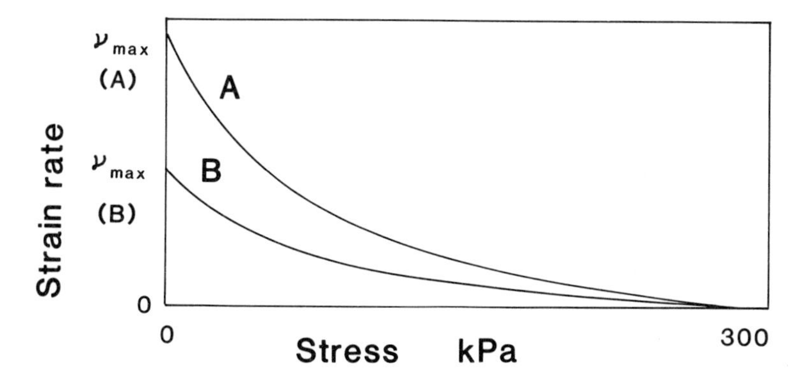

Fig. 7.3. *Hill's equation converted to relate strain rate to stress. Curve A is for a 'faster' muscle than Curve B, but both terminate at the same isometric stress.* ν_{max}, *maximum strain rate.*

of the muscle is taken to be the cross-sectional area of myofibrils, excluding non-contractile components of the muscle cells. This value of stress is also realistic for the whole muscle, in the case of muscles that are adapted for short bursts of anaerobic activity, but not for muscles that are adapted for prolonged aerobic activity. The latter have to contain substantial amounts of mitochondria, which reduces the stress because it increases the cross-sectional area of the muscle without increasing the force exerted (below).

The maximum strain rate (ν_{max}), found by setting $\sigma=0$ in equation (7.10), is

$$\nu_{max} = \beta\sigma_0/a. \tag{7.12}$$

ν_{max} is not constant from one muscle to another of the same general type. This variable, also known as the 'intrinsic speed', differs from one muscle to another, and serves to characterize a particular muscle as 'fast' or 'slow'. At the molecular level, it is related to the rate at which the myosin cross-bridges 'cycle', that is attach, detach, and reattach as they walk their way along the actin filaments. When the muscle is contracting against zero force, this cycling rate determines how long it takes the muscle to shorten by a given fraction of its own length, which is the same as the strain rate.

Adaptation of maximum strain rate

In locomotor muscles, the maximum strain rate has to be adapted in such a way that the muscle can work with a reasonable degree of efficiency. The *conversion efficiency* of a locomotor muscle is the ratio of the mechanical work it produces to the fuel energy it consumes. At the left-hand end of Fig. 7.3 (zero stress), the muscle shortens, but does not exert any force, so the work done is zero. The muscle does consume fuel even under these conditions, therefore the efficiency is zero. The efficiency is also zero if the muscle exerts its maximum stress σ_0, because then it cannot shorten. Once again, it consumes fuel, but does no work. At any value of the stress in between 0 and σ_0, work is done, and the efficiency has some positive value. If the efficiency is plotted against the stress for this range of values, the curve shows a broad peak, with an ill-defined maximum near the middle of the range.

To obtain the maximum amount of work in return for the fuel energy that it expends, the animal has to arrange matters so that its muscles exert an intermediate amount of stress, in the region of 150 kPa, while contracting. This in turn defines the strain rate at which the muscle must shorten. A 'fast' muscle (A in Fig. 7.3) exerts 150 kPa if it shortens at some particular value of the strain rate, whereas a 'slow' muscle (B), forced to shorten at the same strain rate, produces hardly any stress and is therefore inefficient. A much lower strain rate is appropriate for efficient contraction in muscle

B, but at this strain rate the 'faster' muscle A would be inefficient, because the stress would be too high. Both muscles might be adapted to do the same job in locomotion, but muscle B would do it in a larger animal than muscle A. The rate at which the enzyme systems supply energy to the myofibrils has to be set a lower level in muscle B than in muscle A.

According to Pennycuick and Rezende (1984), typical values for the stress and strain prevailing in the myofibrils of bird flight muscles during cruising flight (not maximal exertion) would be 150 kPa and 0.15. The density of vertebrate muscle is about 1060 kg m^{-3}. Therefore the mass-specific work, from equation (7.5), is about

$$Q_m = (1.5 \times 10^5 \times 0.15)/1060 = 21.2\ \mathrm{J\ kg^{-1}}. \quad (7.13)$$

Strain rate and frequency

In cyclic contraction, as discussed above, the muscle must shorten through its active strain in approximately half the time for one complete cycle. The frequency of leg or wing movements is fixed within quite narrow limits, being lower for larger animals of the same general type. The underlying reasons for this were discussed by Hill (1950), but suffice it to say that a condor cannot flap its wings at hummingbird frequencies, because the stress in the muscle attachments needed to produce the required accelerations of the wings, would be far higher than the materials could withstand. From the argument in the last paragraph, this means that the maximum strain rate of the condor's muscles (an adaptation ultimately traceable to the cycling rate of the cross-bridges) *must* be much lower than in the hummingbird, otherwise it will not be able to get any work out of its muscles. This leads to the notion introduced by Pennycuick and Rezende (1984) that any particular locomotor muscle is adapted to work at a well-defined *operating frequency*. The operating frequency determines the mass-specific power output of the myofibrils. From equation (7.7), it is

$$P_m = 21.2f\ \mathrm{W\ kg^{-1}} \quad (7.14)$$

for birds in cruising flight.

Flight at reduced power

A bird that changes from level flight to a shallow descent has to reduce the power output from its flight muscles, and a reduction of both the frequency and amplitude of flapping is often readily visible. Both of these changes have the effect of reducing the strain rate at which the muscles shorten, and probably the stress as well (it is possible to reduce both stress and strain rate simultaneously if the muscle is not maximally stimulated). The argument above shows that if the bird was maintaining a combination of stress and strain rate that resulted in near-maximum efficiency in level flight, then efficiency must suffer if one or both of these quantities are reduced. This may not be important in a short descent to a landing, but long-distance migrants have to reduce power by a large amount in level flight, in circumstances where efficiency is important. Many small passerines set off on migration flights with up to half of their all-up mass consisting of consumable fat. As this fat is used up, V_{mr} progressively declines, and the power required to maintain V_{mr} also declines by a large factor, possibly exceeding two.

The flight muscles of migrants have to be adjusted to run efficiently when producing the maximum power that is required continuously, that is, when the load is heavy, soon after take-off. As the mass declines, power has to be reduced, but if this were done by reducing stress or strain rate during contraction, a loss of efficiency would result, which would lead to a reduction of the migration range. Birds usually solve the problem in a different way, by continuing to run the muscles near their maximum power output, but running them intermittently. Some birds do this by *flap-gliding*. The bird flies level for a while, exerting more power than necessary to maintain its speed, so that it accelerates. Then, it stops flapping and glides for a while, still flying level but slowing down. This mode of flight is common in raptors, pelicans, ibises, shearwaters, albatrosses, and many other medium-sized to large birds. Passerines, and some larger birds, such as woodpeckers, have a different mode of intermittently powered flight, called *bounding*, in which the wings are completely folded up during the non-flapping phase. Level flight is not possible in this mode, because the bird follows a parabolic ballistic trajectory during the non-flapping phase. During the flapping phase, it has to exert a lift force greater than its weight, so as to maintain its speed while following a flight path that curves upwards. The mechanics of bounding flight have been considered theoretically by Rayner (1985) who claims that there are energetic advantages in this type of flight, independent of its function in allowing the bird to reduce power while maintaining constant mechanical conditions in the muscles during the powered phase.

This problem is not confined to birds. Aircraft engines are also most efficient when producing nearly their maximum power output, and suffer a loss of efficiency when 'throttled back'. Intermittent operation consisting of full-power climb followed by a long glide was used in the notorious U2 spy plane, and motor-gliders flown across country in this fashion go much further than if flown level at reduced power. The Rutan Voyager, designed especially to fly non-stop around the world, had two engines. Both were required to run at nearly full power during the initial part of the flight, but later, when some fuel had been consumed, one of them was shut down. It was much more efficient to fly with one engine at full power, and the other one shut down, than to run both engines at half power. One would expect to see the corresponding strategy in migrating passerines, that is, continuous flapping in the initial part of the flight, giving way to bounding, with a progressively decreasing proportion of the time spent in flapping, as the mass declines. This could probably be observed from some of the existing tracking-radar observations of migrating passerines, in which 'radar signatures' indicative of intermittent flapping are seen (Williams *et al.* 1977).

Power output of aerobic muscles

Aerobic muscle

The mass-specific power output estimated by equation (7.14) refers to the myofibrils. It refers to the whole muscle only in the case of anaerobic muscles that consist predominantly of myofibrils, being specialized for short bursts of sprint activity. An aerobic muscle with a substantial mass-specific power output has to contain mitochondria, whose function is to supply ATP to the contractile proteins, at a rate sufficient to sustain their power output. The higher the specific power output (of the myofibrils), the more mitochondria are required to supply energy to each gram of contractile machinery. The effect of this can be seen in a simplified way if the muscle is considered to be made up of two components, myofibrils and mitochondria. Other components, such as nuclei and sarcoplasmic reticulum, are neglected for the moment. The volume (v) of the whole muscle is the sum of the volume of myofibrils (v_c) and the volume of mitochondria (v_t):

$$v = v_c + v_t. \tag{7.15}$$

We suppose that the volume of mitochondria required is directly proportional to the mechanical power output (P):

$$v_t = kP, \tag{7.16}$$

where k is a constant with the dimensions of volume/power. k is the volume of mitochondria required to sustain unit *mechanical* power output, and is referred to as the *inverse power density* of the mitochondria. We now specify that σ is the stress exerted across the myofibrils (not across the muscle as a whole), in which case, from equation (7.6),

$$P = \sigma\lambda f v_c. \tag{7.17}$$

We can now find the volume-specific power output for the whole muscle, which is

$$\begin{aligned} P_v &= P/(v_c + v_t) \\ &= \sigma\lambda f v_c/(v_c + k\sigma\lambda f v_c) \\ &= \sigma\lambda f/(1 + k\sigma\lambda f). \end{aligned} \tag{7.18}$$

The mass-specific form of this is

$$P_{\mathrm{m}} = \sigma\lambda f/\rho(1+k\sigma\lambda f). \tag{7.19}$$

Comparison of this result with equation (7.7) shows that the effect of adding mitochondria is to divide the specific power output of the myofibrils alone by a factor $(1+k\sigma\lambda f)$. At very low operating frequencies (below about 5 Hz), this factor is negligibly greater than 1, so that equation (7.7) can be considered to apply to the whole muscle. However, at very high frequencies, $k\sigma\lambda f$ is much greater than 1, and in that case, equation (7.19) approximates to

$$P_{\mathrm{m}} = 1/\rho k. \tag{7.20}$$

Rather than increasing indefinitely at very high frequencies, as equation (7.7) would suggest, the specific power output tends towards an upper limit, determined only by the properties of the mitochondria, and independent of the mechanical properties of the myofibrils. This theoretical limit represents a muscle that consists entirely of mitochondria, except for an infinitesimal amount of myofibrils, contracting at an infinite frequency. Of course, no real muscle approaches close to this limit, although the flight muscles of hummingbirds and some insects contain approximately equal volumes of myofibrils and mitochondria.

Scaling of power available and power required

'Scaling' is a form of reasoning that predicts how a particular quantity should vary, if the size of the animal is changed, while preserving a constant shape. For instance, if one imagines a series of 'geometrically similar' birds of various sizes, both the mass and the volume should vary with the cube of the length. This can be written in a form of shorthand as $m \propto l^3$ and $v \propto l^3$, or conversely as $l \propto m^{1/3}$.

This has been discussed at greater length by Pennycuick (1975), who notes that flapping frequency would be expected to scale with a different power of the mass, depending whether the bird is in cruising flight, or exerting maximum effort. The suggested values are $f \propto m^{-1/3}$ for maximum exertion, and $f \propto m^{-1/6}$ for cruising. According to equation (7.7), the mass-specific power is obtained by multiplying the wingbeat frequency by the stress and the strain, neither of which is expected to vary with the mass. Therefore the mass-specific power should scale in the same manner as the frequency, and the absolute power available from the muscles should vary with some power of the mass between $\frac{2}{3}$ and $\frac{5}{6}$, in other words, in much the same manner as basal metabolic rate, and other forms of energy consumption.

However, the power required to fly at some characteristic speed, such as the minimum power speed, scales with a higher power of the mass. The argument given by Pennycuick (1975) indicates that it should scale with the $\frac{7}{6}$ power of the mass. Real birds deviate from geometrical similarity in a manner that would tend to make the exponent less than $\frac{7}{6}$, perhaps as low as 1. However, no amount of allometry can altogether escape the trend for power required to increase more steeply than power available, as the mass increases. The trend has readily visible consequences, in that small birds (especially hummingbirds) have power to spare for energetically demanding manœuvres such as hovering, low-speed flight, and jump take-offs. Among very large birds, only swans regularly migrate for substantial distances in flapping flight. The others depend to varying degrees on soaring, condors to such a degree that they appear to have little power to spare over that needed to fly level at their minimum power speed.

Upward extrapolation of this trend indicates that there must be an upper limit to the mass of an animal that can fly by muscle power. It is difficult to establish exactly what the heaviest living bird is, that is able to fly horizontally, but records suggest that the upper limit is between 12 and 16 kg. On the other hand some fossil birds and pterosaurs are known that were considerably larger than any living bird. Perhaps no living bird is actually on the upper limit, or perhaps larger flying animals were able to exploit some reliable form of soaring, that permitted them to fly even though they were not able to sustain horizontal flight.

A second conclusion that may be noted is that the power that birds have to exert in order to fly (that is, the power required) scales at a steeper slope than the basal metabolism. Therefore if the power consumption is expressed as a multiple of the basal metabolic rate (a procedure which physiologists often use but never justify), then this multiple must be larger in large birds than in small ones. Even if any such quantity as 'flight metabolism' could be identified, it could not conceivably be a constant multiple of the basal metabolic rate. This point is illustrated in example 7 of Chapter 8.

Cost of maintaining a steady force

A muscle that exerts a steady force, but does not shorten, does no external work, although it does consume fuel energy. The 'efficiency' as defined above is, of course, zero, but this is hardly relevant if the production of work is not the muscle's function. A different measure is needed, that expresses the power needed to maintain a given amount of force. A physiologist would respond with a measurement expressed in some such units as calories per hour per gram-force, but as usual this approach obscures the nature of the problem and makes it appear more complicated than it is. It is better to consider the dimensions of power (ML^2T^{-3}) and of force (MLT^{-2}), and to divide one by the other. This indicates that the required quantity has the dimensions of speed (LT^{-1}). If the measurement is to characterize the muscle type, rather than the individual muscle, we need to divide the volume-specific power (dimensions $ML^{-1}T^{-3}$) by the stress (dimensions $ML^{-1}T^{-2}$). In this case the measure of the cost of maintaining stress has the dimensions T^{-1}.

It was noted above that the 'speed' of a muscle may be characterized by a quantity with these same dimensions, the maximum strain rate, or 'intrinsic speed'. This in turn is related to the cycling rate of the myosin cross-bridges (actually a frequency, with the same dimensions). It has been known for many years that a 'slow' muscle, that is one with a low maximum strain rate, expends less power than a 'fast' muscle when maintaining the same isometric force. There is abundant physiological evidence to this effect, but to judge from such reviews as that of Johnston (1985), there do not seem to be any empirical data that would allow the connection between the energetic cost of maintaining tension and the maximum strain rate of the muscle to be quantitatively established.

From inspection of the dimensions, one might surmise that if a muscle whose maximum strain rate is ν_{max} expends fuel energy at a volume-specific rate P_v when exerting an isometric stress σ, then the dimensionless ratio $P_v/\sigma\nu_{max}$ would have a constant value for different muscles, and over a range of stresses. The equations given by McMahon (1984) supply a value for this ratio for the case where the muscle is exerting the maximum isometric stress, and P is the rate at which energy is liberated by splitting ATP molecules (rather than by oxidizing fuel). In this case $P_v/\sigma = \nu_{max}/16$.

Scaling of power requirements for gliding

Gliding is not 'effortless' for a bird as some people imagine. The reader can verify this if he will support his weight on his elbows for a few hours in between two filing cabinets. A gliding bird has to support its weight in the same way, holding its wings level by exerting a constant force in its pectoralis muscles (Fig. 7.4). If the centre of lift on the wing is a distance j_1 from the shoulder joint (the *moment arm*), then the *moment* (M) exerted by the pectoralis of one side about the shoulder joint is

$$M = mgj_1/2. \tag{7.21}$$

This moment has to be balanced by a force (F) exerted by the pectoralis muscle, whose moment arm about the shoulder joint is j_2 (Fig. 7.4). In a series of geometrically similar birds, j_1 (being a length) would scale with $m^{1/3}$, and therefore the moment (M) that the muscle is required to exert would scale with $m^{4/3}$. If j_2 also scales with $m^{1/3}$, then the force that the

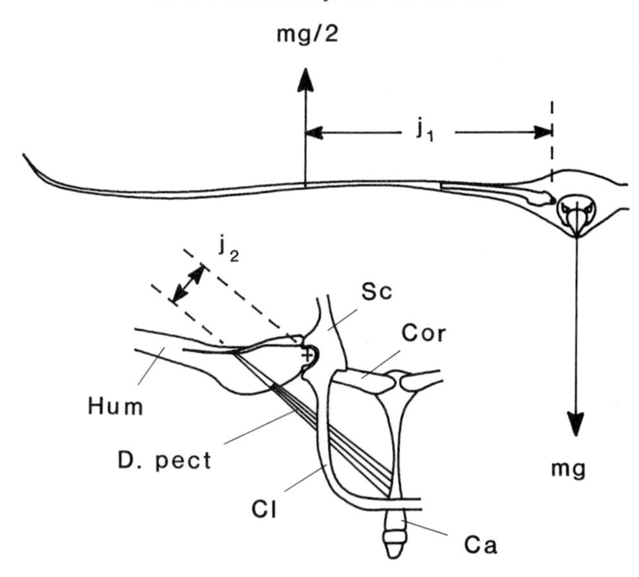

Fig. 7.4. *Each wing of a gliding bird has to support half the weight, acting through some point at a distance* j_1 *from the shoulder joint. The moment has to be balanced by the pectoralis muscle, or part of it, exerting a larger force with a smaller moment arm* (j_2). *Old-world vultures (below, left) have a separate deep portion of the pectoralis (D. pect), which is thought to be a specialized tonic muscle for this function. Hum, humerus; Cor, coracoid; Sc, scapula; Cl, clavicle; Ca, carina of sternum. From Pennycuick (1972).*

muscle is *required* to exert would be directly proportional to the mass, but the force that it is *able* to exert, if the stress is constant, would be proportional to its cross-sectional area, and hence to $m^{2/3}$. Clearly some allometry is inescapable. If the moment arm j_2 were indeed proportional to $m^{1/3}$, then the muscle would have to scale allometrically, so that its cross-sectional area scaled with m^1 instead of $m^{2/3}$. Alternatively, the allometry could be in the moment arm, or partly in the muscle and partly in the moment arm.

As there is no empirical information on this allometry, we can assume for the sake of a simple argument that the first of the above alternatives is correct, and that the force exerted by the muscle is directly proportional to the mass. The volume-specific (and also mass-specific) power required by the muscle to exert this force is expected to be proportional to the maximum strain rate (ν_{max}, above). There is no information about this either, but it is reasonable to assume that it would be slower in larger birds than in smaller ones, probably scaling in the same way as the maximum strain rate of the locomotor muscles, that is somewhere between $m^{-1/3}$ and $m^{-1/6}$. If we assume that the mass of the muscle supplying the force is a constant fraction of the mass of the whole bird (roughly true for the pectoralis as a whole, according to Greenewalt 1962), then the power required to maintain the force (P_F) is proportional to both the body mass and the maximum strain rate:

$$P_F \propto m\nu_{max}. \tag{7.22}$$

Thus the power required would scale somewhere between $m^{2/3}$ and $m^{5/6}$. This is approximately the same law as applies to the scaling of basal metabolic rate. Therefore it is not unreasonable to assume that the power required to hold the wings level is a fixed multiple of the basal metabolic rate in birds of different size. This conclusion contrasts with the scaling of the power required for flapping flight. As noted above, the latter must scale with a higher power of the mass than the basal metabolic rate, so that it is not conceivable that the total power required could be a constant multiple of the basal metabolic rate. In flapping flight, this multiple has to be less in large birds than in small ones, but not so for gliding flight.

There is one published result on a gull gliding in a wind tunnel, in which Baudinette and Schmidt-Nielsen (1974) found that the total power required was about twice the basal metabolic rate. The above argument suggests that it is not unreasonable to apply the same ratio to other gliding birds, and Program 2 does this, there being no empirical information. Admittedly the justification for this extrapolation is weak, but it is better than no justification at all, which is the usual basis of such generalizations in physiology.

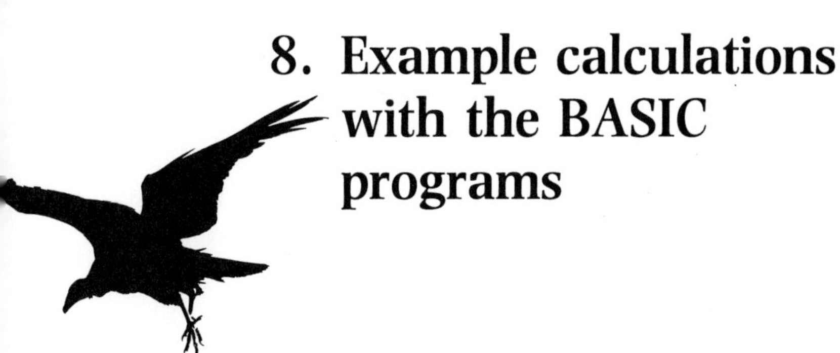

8. Example calculations with the BASIC programs

Introduction

Some of the examples in this chapter are identified as test examples, which have been provided so that readers who have typed in their own copies of the programs can test them for correct operation. It is essential to run the examples, and check that the output agrees exactly with the specimen output reproduced in Tables A1.1 to A1.9 in Appendix 1. Any discrepancies, either in the numbers or the layout of the tables, will be due to errors in entering the programs, or to syntax differences between different versions of BASIC. Advice on dealing with such problems is in Chapter 4. The programs should not be used for original research until the test examples have been run with no errors or discrepancies. It is also best to make sure the programs work in their original form, before attempting any modifications.

All of the following examples use real data. The reader will find it instructive to work through the examples, even if using a prerecorded program disc. They have been designed to illustrate the various features of the programs, and also to show how the programs can be applied to problems of biological interest. Some of them illustrate ways in which the output of the programs can be compared with empirical observations. Ultimately, confidence in the predictions of these programs will depend on future users' ingenuity in devising ways to make such comparisons.

Checking Program 1 against physiological results

In theory it should be possible to measure the rate at which a bird in flight consumes fuel energy, and to compare the result with the predictions of Program 1. There are indeed large numbers of estimates of 'flight metabolism' in the literature, based on measurements of oxygen consumption, on carbon dioxide production determined by stable isotope methods, or on food intake. Estimates of 'flight metabolism' usually refer to measurements of the total amount of energy consumed while the bird flew around for a few hours at an unknown speed, climbing or descending at will, and performing occasional energetic manœuvres such as landings and take-offs. Such crude measurements have uses in ecology, to provide a rough empirical measure of the amount of fuel that a bird consumes in the course of a typical day's activity, but they cannot be used for comparison with predictions made from any mechanical theory. To perform multivariate regressions of such estimates against various body measurements, as some authors have done, is absolutely meaningless.

For an experimental result to be compared with predictions from Program 1, the bird has to fly horizontally at a steady (and known) speed for a period that is long enough to allow it to settle down into steady, aerobic, cruising respiration. Then, observed values have to be supplied for the following five variables:

(1) bird's all-up mass, optionally divided into empty mass and payload;
(2) wing span;
(3) airspeed;
(4) air density;
(5) measured power.

Claims that results have been compared with various 'models' may be discounted as worthless, if the necessary measurements are not supplied. In only one published study, that of Rothe, Biesel, and Nachtigall (1987) on the oxygen consumption of pigeons in prolonged flight in a wind tunnel, are the conditions of flight well enough defined, and the required data explicitly enough presented, for the comparison to be made. Even in this case, the air density was not recorded, and the measurements of power output were presented in a form that made it impossible to reconstruct the original observations.

Example 1. Oxygen consumption of the pigeon

(Test for Program 1: output in Tables A1.1 to A1.3.)
The first example has been selected partly to show sceptical readers that Program 1 actually does estimate a bird's energy consumption in flight. Rothe and Nachtigall (1987) undertook lengthy experiments to find a breed of pigeon that would fly steadily for prolonged periods in a wind tunnel, and to establish the conditions under which various physiological measurements could be reliably obtained. On this basis, Rothe, Biesel, and Nachtigall (1987) made oxygen consumption measurements on several pigeons flying at different speeds. The basis for comparison is their Fig. 5, in which fourteen power estimates, derived from oxygen consumption measurements on three different pigeons flying at various speeds, were plotted against airspeed. The power is expressed in the form of 'specific metabolic rate', that is, power

per unit body mass which, regrettably, is not the form in which the data are required. To convert the data back into the original observations of power, the body mass of each pigeon on each occasion would be needed, but this information is not supplied. The pigeons were, however, divided into three categories, 'heavy', with a mean mass around 0.35 kg, 'medium' (0.33 kg), and 'light' (0.31 kg). As a reasonable compromise, all of the 'specific metabolic rate' observations have been multiplied by 0.33 kg, to convert them back into powers. Wing span is mentioned as '60 cm', so it has to be assumed that this value applied to all the different pigeons. Air density was not recorded. Saarbrücken, West Germany, where the experiments were carried out, is situated at about 320 m above sea level, where the density in the standard atmosphere would be 1.19 kg m^{-3}. This value has been assumed, although the actual air density doubtless varied from day to day.

The data needed for input to the program (mass, wing span, and air density) are shown at the head of each output table. To get the output as shown in Table A1.1, run Program 1 and, when prompted for data, enter 0.31 kg for the body mass, 0 for the payload mass, and 0.6 m for the wing span. Before proceeding, the air density must be changed to 1.19 kg m^{-3}, by selecting Number 10 from the menu. Then, select Number 14 to run the program. At the prompt 'Version', enter 'light'. The computer should respond by printing out a table identical to the one reproduced as Table A1.1. When invited to 'Do more', respond 'Y'. The menu will reappear, with the same body measurements and air density that you entered before. You now choose number 2 from the menu, and change the empty mass to 0.35 kg. Notice that when the menu reappears, the body cross-sectional area has been recalculated for the new body mass. Choose Number 14 to run the program again, and enter 'heavy' for 'Version'. This time the output should duplicate Table A1.2. The message at the top of the first page of output is not repeated on subsequent runs. For the third run, change the head-wind strength (initially zero) to 8 m s^{-1}, and get Table A1.3. Notice that the fat consumption (per kilometre over the ground) is dramatically increased by this, and also that no fat consumption estimates are printed for airspeeds below 9 m s^{-1}, because the pigeon would make no progress, or be blown backwards at those airspeeds.

Figure 8.1 was created by a graph-plotting version of Program 1, and shows calculated power curves for 'heavy', 'medium', and 'light' pigeons, with Rothe, Biesel, and Nachtigall's 14 observations superimposed. Most of the observations fall between the 'medium' and 'heavy' curves, with a slight tendency for the experimental values to be higher at the lowest speeds. The lone observation at 8.2 m s^{-1} was contributed by the only pigeon that would fly at such a low speed. The authors commented on the reluctance of the pigeons to fly at very low speeds, and this is associated

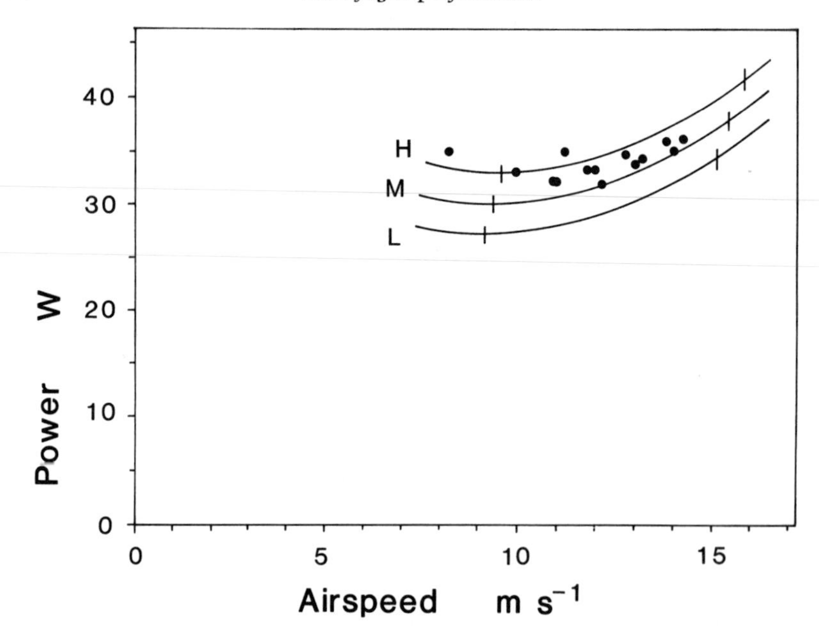

Fig. 8.1. *Output of a graph-plotting version of Program 1, for pigeons with 0.6 m wing span and three values of the body mass: H, heavy (0.35 kg); M, medium (0.33 kg); L, light (0.31 kg). The points are from oxygen consumption measurements on pigeons flying in a wind tunnel by Rothe, Biesel, and Nachtigall (1987). The vertical marks on the curves are calculated values of* V_{mp} *and* V_{mr}.

with observations of power consumption that were higher, and also less consistent, than those obtained at higher speeds. This would be readily understandable in terms of Rayner's (1986) idea of 'gaits', if the pigeons were forced to shift from a concertina gait to a single-ring gait, when made to slow down to the region of the minimum power speed (see Chapter 5). If this is the correct interpretation, the pigeons should also show a change of wing action as they shift to the lower-speed gait.

The curves plotted in Fig. 8.1 are estimates of 'chemical' power, that is, they have been obtained from the calculated mechanical power, by dividing it by the conversion efficiency (η). A small adjustment to the assumed value of η (0.23) would move the calculated curves so that they coincide with the observed points. Adjustment of some other variables would also effect this, for example the circulation and respiration factor (*R* in the program). However, the program should not be too hastily amended as the small discrepancy between the points and the curves is not necessarily due to an error in the calculation. It is just as likely to be due to one or more of the many sources of error that are inherent in physiological experiments, or to the conversion of the experimental observations from rate of oxygen consumption into power.

Other problems involving fuel consumption—Program 1

Example 2. Migration range of the snow goose

(Test for Program 1, including effect of radio, and migration range. Output in Tables A1.4 to A1.7.)

The second example (snow goose) is included to illustrate the use of the subsidiary program that finds the range of a long-distance migrant, and also the effect of strapping a rather large radio transmitter to the bird's back. The numbers used for body mass and payload (here equated with consumable fat rather than crop contents) are taken from field observations by Gauthier *et al.* (1984). The full exercise involves six runs of the main program, and three of the subsidiary range program. The output of two runs of the main program is reproduced as Tables A1.4 and A1.6, and that of three runs of the range program in Tables A1.5 and A1.7.

First, enter 2.59 kg for the empty mass, 0.65 kg for the payload, and 1.6 m for the wing span. When the main menu comes up, change the air density to 1 kg m^{-3}, to represent a flying height of about 2000 m ASL, and then run the main program. For 'Version' put 'Full fuel no radio' (no commas). The output (Table A1.4) represents the flight performance of the goose as it sets out on its migration with a full load of fat. The number you want from Table A1.4 is the maximum L/D (8.88). Now return to the menu and change the payload to zero (fuel used up). Run the main program again (output not reproduced). This time L/D max is 9.47. The increase is due to the body slimming down as fuel is consumed, and thus presenting a smaller frontal area, and creating less drag. Take the average of these two figures (9.18) as an estimate of the average effective lift:drag ratio for the whole flight.

Now invoke the range program by selecting Number 15 from the main menu. It asks you if you want to go back to the main menu to find the lift:drag ratio. You can respond 'N', as you have already done this. Enter 9.18 when prompted for the lift:drag ratio, and 0.65 kg for the fat load at take-off. You now get a subsidiary menu showing the values you have just entered. Choose Number 4 to get the output shown in Table A1.5. The range (1880 km) is the *air* distance the bird can fly to 'dry tanks' (not adjusted for wind). Note that all numbers in the output tables are rounded to three significant digits, so the range is rounded to the nearest 10 km. The subsidiary menu reappears, and you can repeat the range calculation any number of times with any changes you want in the data. When finished, choose Number 5 to return to the main menu. Changes in the empty mass and payload, made through the subsidiary menu, are not carried over when you go back to the main menu. Instead, the program remembers the values that were in force before you diverted to the range program.

Now equip the snow goose with an early-model satellite-trackable radio transmitter, as used in wind tunnel tests by Obrecht, Pennycuick, and Fuller (1988). The plain version of this formidable device (Type 3A) had a mass of 160 g. Its frontal area was 2130 mm^2, and it was found to increase the equivalent flat-plate area of the snow goose body by 1280 mm^2. Choose number 16 from the menu to add the radio. When prompted for the mass, enter 0.16 kg (not 160). The equivalent flat-plate area added by the radio is 0.00128 m^2. An easier way to enter this small number is to note that to convert 1280 mm^2 to m^2 you have to multiply by 10^{-6}, so enter 1280E−6 (see Chapter 4). As a rule-of-thumb, the equivalent flat-plate area of a radio transmitter is about half the actual frontal area, if the transmitter is fully exposed on the bird's back, and has streamlined end fairings as described by Obrecht, Pennycuick, and Fuller (1988). If the transmitter is completely covered by the feathers, for instance under the ventral tail coverts, you may be justified in calling its equivalent flat-plate area zero.

Having installed the radio, do two more runs of the main program, with the payload set to 0.65 kg and zero as before. The output (Table A1.6) has two additional lines, showing the mass and flat-plate area that you entered for the radio. The estimates for the maximum L/D are now 8.24 and 7.84 (mean 8.04). The output of the range program (first block in Table A1.7) has an additional line for the radio mass, and the range is cut to 1560 km. The range is reduced partly because of the lower effective lift:drag ratio, and partly because the initial fuel load (0.65 kg) is a smaller fraction than before of the all-up mass at take-off (equation (3.15)).

Back at the main menu, select Number 16 again and enter 0.16 kg as before for the radio mass, but this time enter zero for the flat-plate area. The goose is now equipped with a radio that is as heavy as before, but dragless. Run the main program twice, as before. The maximum L/D actually increases slightly as compared to the first pair of runs (8.89 and 9.50, mean 9.20), but the range is less (1780 km), because the fuel load at take-off (0.65 kg) is now a smaller fraction of the all-up mass. Note that if the radio has both mass and drag, the range is reduced by 17 per cent, but if the radio has the same mass but no drag, the reduction is only 5 per cent.

Example 3. White-tailed tropicbird

Schaffner (1988) characterizes the white-tailed tropicbird as a pelagic bird, which forages at sea far from the nest, and brings back relatively large loads of food for its young, at rather infrequent intervals. A typical empty body mass is 0.37 kg and wing span 0.92 m. If there is no payload, Program 1 estimates V_{mr} to be 13.0 m s^{-1}. If the bird is carrying 50 g in its crop (a typical payload when feeding young), V_{mr} is 13.5 m s^{-1}.

The output is not reproduced in full for this example, as the program features have been tested by Examples 1 and 2. The reader may like to use Program 1 to duplicate the results in Table 8.1. Air density can be left at

Table 8.1. White-tailed tropicbird. Round-trip time and cost for a 50 km flight.

Wind	Flight time		Fat consumed
m s^{-1}	h	min	g
0	2	08	5.29
6.8 (upwind outbound)	2	56	7.01
6.8 (downwind outbound)	2	56	7.56

Fig. 8.2. *White-tailed tropicbirds display flying. These pelagic birds feed mainly on flying fish and squid, which they catch by plunge diving. Puerto Rico.*

the default value (sea level) for this example. The tropicbird makes a hypothetical foraging flight to a feeding area 50 km from the nest, carrying no payload outbound, and 50 g on the way back. For simplicity, the airspeed is assumed to be 13.0 m s^{-1} in both directions, regardless of wind. If there is zero wind, the flight time is 1 h 4 min in each direction, i.e. 2 h 8 min for the round trip. The fat consumption in g km^{-1} is given opposite 13 m s^{-1} in the fifth column of the output table, and is multiplied by 50 to give the amount of fat used on either the outbound or the inbound leg. We now impose a steady wind of 6.8 m s^{-1}, which is typical for the trade wind zones in which tropicbirds live, and run the program twice more, once with a head wind of this strength, and then with a tail wind. The effect of this is to increase the flight time for the round trip by 38 per cent, regardless of whether the tropicbird goes upwind to the foraging area and returns downwind, or vice versa. However, the fat consumed increases by 33 per cent if the outbound leg is upwind, and by 43 per cent if the outbound leg is downwind. The reader will be able to see the reason for this asymmetry by studying the output tables. The conclusion is that if a number of islands are available as nesting sites, the tropicbird should select one that is to leeward of the feeding area, other things being equal.

Problems involving mechanical power only—Program 1A

Example 4. Double-crested cormorant

(Test example for Program 1A: output in Table A1.8)
The 'span-ratio method' is a more direct (because strictly mechanical) method for checking the calculations than the physiological method of example 1. It is also much simpler, and can be applied in the field. It begins from the assumption that a bird in cruising flight is generating a concertina wake, with equal circulation in the upstroke and downstroke, as observed by Spedding (1987) in the kestrel (see Chapter 5), and also that the spanwise lift distribution is the same in the upstroke and the downstroke. If this is so, then the ratio of the lift on the upstroke to that on the downstroke must be the same as the *span ratio*, that is the ratio of the wing span on the upstroke to that on the downstroke. The span ratio can be measured in the field from video recordings. Combined with measurements of airspeed, it yields an estimate of the effective lift:drag ratio as defined in equation (3.13). The details of the practical calculation, and the assumptions involved, are given by Pennycuick (1989). Program 1A rather than Program 1 is used to calculate the effective lift:drag ratio for comparison with these observations, because the power calculation has to include mechanical

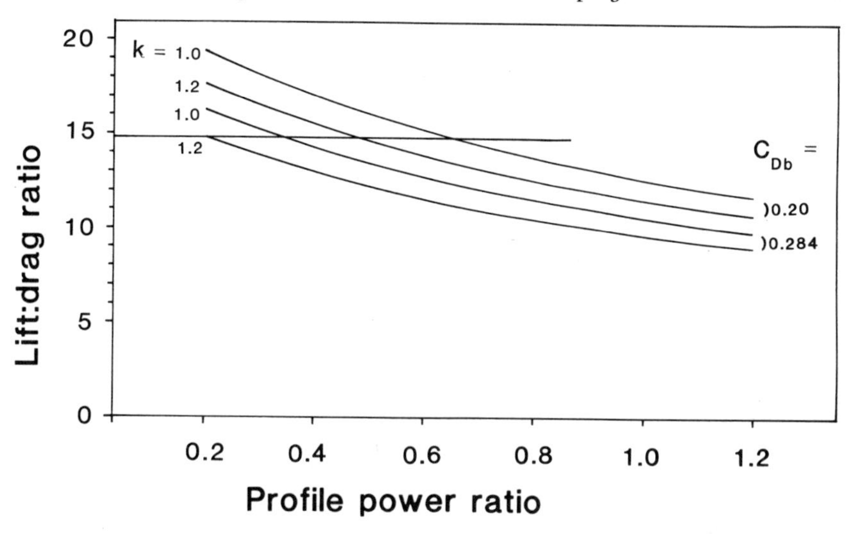

Fig. 8.3. *Results of 24 runs of Program 1A on the double-crested cormorant, showing the effect on the maximum lift:drag ratio of systematically varying the profile power ratio* (X_1), *the induced power factor* (k), *and the body drag coefficient* (C_{Db}). *The horizontal line is a field estimate of lift:drag ratio made by the span-ratio method (Pennycuick 1989).*

components of power only. It is not concerned with basal metabolism, or with the conversion of fuel energy into work.

Although not in principle confined to steady (or even level) flight, the span-ratio method is easiest to apply to birds that cruise at a steady speed, using continuous flapping flight (not flap-gliding). To date it has been tried on the double-crested cormorant by Pennycuick (1989), for which it yielded estimates for N_{max} around 14. Program 1A, when run with default values for the variables, gave a much lower estimate (near 9) for the same quantity (Table A1.8). The graphs of Fig. 8.3 are designed to show how Program 1A can be used to find out whether the discrepancy between the field observation and the output of Table A1.8 can be eliminated by adjusting the values of variables. Program 1A was run 24 times, systematically varying the values of the body drag coefficient (C_{Db}), the induced power factor (k), and the profile power ratio (X_1). This is the basis of the statements in Chapter 5, to the effect that some evidence suggests that the assumed default values for these variables may be too high. Of course, the discrepancy could also be wholly or partly due to some source of bias in the span-ratio method. More experience is needed with methods of this type before any firm conclusion can be drawn.

Fig. 8.4. *Giant petrel, a typical pelagic soaring bird, with pointed wings and an aspect ratio of about 12. South Georgia.*

Gliding and soaring—Program 2

Example 5. Soaring performance of vulture versus giant petrel

(Test example for Program 2: output in Table A1.9)
The African white-backed vulture is typical of large birds that are specialized for soaring in thermals over land. Giant petrels, although somewhat vulturine in their habits, get most of their energy by skimming along the windward slopes of ocean waves. Giant petrels never encounter thermals, and certainly would not know how to soar in them if they did, but this need not restrain the reader from testing their performance at thermal soaring with the aid of Program 2. The complete output for the white-backed vulture is reproduced in Table A1.9, and may be used for testing correct operation of Program 2. The body measurements of both birds are in Table 8.2, together with salient points from the output, for comparing their performance. Both were tested at sea-level air density, so as to make the results comparable.

Fig. 8.5. *Rüppell's griffon vulture, a larger but closely related congener of the white-backed vulture used for Example 5. The aspect ratio is about 7, and the primaries are deeply emarginated, giving prominently slotted wing tips. This wing shape is typical of birds that soar in thermals over land. Northern Tanzania.*

Both birds have about the same mass, but the petrel has slightly less wing span than the vulture, and less than half the wing area. This means that its wings are narrower, having a higher aspect ratio (Figs. 8.4 and 8.5). The petrel's wing loading is almost exactly twice that of the vulture,

Table 8.2. White-backed vulture compared with giant petrel. Estimates from Program 2.

	Vulture	Giant petrel
Mass (kg)	5.38	5.19
Wing span (m)	2.18	1.99
Wing area (m^2)	0.690	0.331
Stall speed ($m\ s^{-1}$)	8.82	12.5
Minimum sink ($m\ s^{-1}$)	0.816	0.834
Best glide ratio	14.6	16.2
Circling radius (m)	22.3	44.7
Cross-country speed in 3 $m\ s^{-1}$ thermals ($m\ s^{-1}$)	12.8	14.2

which means that its stalling speed is higher by a factor of $\sqrt{2}$. However, its minimum sink speed is only slightly higher, and its best glide ratio is 11 per cent better. The petrel's circling radius is twice that of the vulture, other things being equal, and this would adversely affect its climbing performance in narrow thermals. On the other hand, thermals are often wide enough to support gliders with still greater circling radii. In conditions that would allow both birds to climb at 3 m s^{-1}, the petrel would be capable of 11 per cent higher cross-country speed. The vulture's extremely broad wing is not ideal for any aspect of soaring except climbing in very narrow thermals. One probably should look at other aspects of performance, such as take-off, to account for it.

Example 6. To flap or not to flap? Common crane versus ovenbird

For birds that breed in northern Europe, the migration route to Africa is barred by the east–west Mediterranean Sea. The Mediterranean is a poor area for soaring, and most soaring migrants make long detours to avoid flying over it, either east via the Bosphorus or west via Gibraltar. These storks and raptors migrate only by day, and never resort to flapping if there is any way to avoid it. The common crane is different. Cranes cross the Mediterranean at night in flapping flight. Further north, they soar like storks if thermal conditions are good, but readily resort to level flapping flight if not.

For a bird that has a choice whether to flap or soar, the tradeoff is that a given distance, if covered by soaring rather than flapping, requires less energy but more time. The reader may like to investigate this by running Programs 1 and 2, using the body measurements in Table 8.3. Both programs were run with an air density of 1.00 kg m^{-3}, representing a typical flying height in the region of 2000 m ASL. In strong thermals (3 m s^{-1} climbs), the crane can achieve 59 per cent of its maximum range speed, with only 9 per cent of the fuel consumption. If the thermals are weak (1 m s^{-1} climbs), the cross-country speed is only 35 per cent of V_{mr}, and the fuel consumption is 15 per cent of that for flapping flight. Cranes will usually attempt to soar even under these conditions, but they also have an intermediate mode of migrating in which they flap intermittently on the inter-thermal glides, so stretching the glides by a variable amount. The theory of this is discussed by Pennycuick, Alerstam, and Larsson (1979). In terms of speed and cost this mode of 'soaring with powered glides' is intermediate between pure flapping and pure soaring.

Table 8.3. Flapping performance (from Program 1) compared with gliding performance (Program 2) for common crane and ovenbird.

	Crane	Ovenbird
Mass (kg)	5.50	0.0174
Wing span (m)	2.40	0.230
Wing area (m^2)	0.720	0.0101
Flapping:		
Maximum range speed ($m\ s^{-1}$)	23.4	10.8
Fat consumption ($g\ km^{-1}$)	0.542	0.00416
Soaring in 1 $m\ s^{-1}$ thermals:		
Cross-country speed ($m\ s^{-1}$)	8.12	3.86
Fat consumption ($g\ km^{-1}$)	0.0821	0.00442
Soaring in 2 $m\ s^{-1}$ thermals:		
Cross-country speed ($m\ s^{-1}$)	11.3	5.39
Fat consumption ($g\ km^{-1}$)	0.0587	0.00316
Soaring in 3 $m\ s^{-1}$ thermals:		
Cross-country speed ($m\ s^{-1}$)	13.7	6.47
Fat consumption ($g\ km^{-1}$)	0.0486	0.00264

Soaring, as every birdwatcher knows, is characteristic of medium-sized and large birds such as hawks, gulls, crows, pelicans, etc. Small passerines are not seen gliding in careful circles in thermals. Many of them make lengthy migrations, so why do they not take advantage of the energy freely available in the atmosphere? The reason can be seen from Table 8.3, which includes the results of running Programs 1 and 2 for an ovenbird, a common passerine north–south migrant from the New World. Its cross-country speed with 3 $m\ s^{-1}$ climbs in thermals (if it were to engage in such uncharacteristic behaviour) would be about 60 per cent of its V_{mr}, and 36 per cent with 1 $m\ s^{-1}$ climbs. These percentages are about the same as for the crane, although the actual speeds are less than half as great. The energetic advantage offsetting the loss of speed is less, or non-existent, in the ovenbird. It requires 63 per cent as much energy per kilometre when soaring in 3 $m\ s^{-1}$ thermals as when flapping at V_{mr} (compared to 9 per cent for the crane). With 1 $m\ s^{-1}$ climbs it actually requires 6 per cent *more* energy for each kilometre flown. This is because basal metabolism is a much larger fraction of the total energy requirement in a small bird than in a large one. The reduction of speed means that the bird spends nearly three times as long covering each kilometre, and during that extra time it is burning up fuel for basal metabolism. This more than offsets the saving of energy due to gliding instead of flapping.

The reader can see where the energy is going by studying the output tables produced by running Programs 1 and 2 with the data of Table 8.3. Remember to set 'Type of bird' to 'passerine' for the ovenbird (so as to get the right version of Lasiewski and Dawson's equations), and change the air density to 1 kg m^{-3}. The next example will throw further light on the varying importance of basal metabolism in birds of different shapes and sizes.

Scale effects with Program 1

Example 7. Twenty-four birds from my files

As noted in Chapter 1, 'flight metabolism' is not a quantity that can be defined with any meaningful degree of precision, but whatever it is, it cannot possibly be a fixed multiple of the basal metabolic rate in a series of birds of different shapes and sizes. To illustrate the point, I have run Program 1 for 24 birds from my files of body and wing measurements, from a yellowthroat (10.3 g) to a tundra swan (9.5 kg). For each bird I have taken P_{mr}, the total power required to fly at the maximum range speed (including basal metabolism) as being most likely to correspond to the kind of observation physiologists hope to get when they measure 'flight metabolism'. Since P_{mr} is expressed in the output table in mechanical form, I divide it by P_{met}, the mechanical equivalent of the basal metabolic rate, as estimated from Lasiewski and Dawson's (1967) regression equations.

The ratio P_{mr}/P_{met} (listed in Table 8.4) varies from 4.86 for the yellowthroat to 64.4 for the tundra swan, a range of over 13:1. The points are plotted in Fig. 8.6 with a reduced major axis line, of which the slope is 0.35, indicating that the ratio P_{mr}/P_{met} varies with the 0.35 power of the mass for this particular data set. The reasons for this strong trend, and also for the large amount of scatter about the line, will be easily understood by the reader who has studied Chapter 3. The practical conclusion is that you cannot get a meaningful estimate of 'flight metabolism' by estimating the basal metabolic rate from Lasiewski and Dawson's equations, and then multiplying by a fixed factor. This simply does not work.

Further uses of the programs

Having worked through the examples, readers will be ready to use the programs on their own problems. Some users will want estimates of the amount of energy that birds use going about their errands, while others

Table 8.4. Cruise power and basal metabolism. Power estimates from Program 1.

	Mass	Span	P_{mr}/P_{met}
	kg	m	
Yellowthroat *Geothlypis trichas*	0.0103	0.167	4.86
Swainson's thrush *Catharus ustulatus*	0.0320	0.285	5.86
American robin *Turdus migratorius*	0.0837	0.400	7.95
Ovenbird *Seiurus aurocapillus*	0.0174	0.230	4.89
Cedar waxwing *Bombycilla cedrorum*	0.0358	0.286	6.32
Yellow-bellied sapsucker *Sphyrapicus varius*	0.0446	0.396	7.45
Laughing gull *Larus atricilla*	0.288	1.00	9.92
Mourning dove *Zenaida macroura*	0.102	0.430	12.9
Wandering albatross *Diomedea exulans*	8.73	3.03	37.3
White-chinned petrel *Procellaria aequinoctialis*	1.37	1.40	22.3
Cape pigeon *Daption capense*	0.433	0.875	16.3
Wilson's storm-petrel *Oceanites oceanicus*	0.0380	0.393	6.66
Sooty tern *Sterna fuscata*	0.195	0.875	8.69
Brown pelican *Pelecanus occidentalis*	3.30	2.39	22.0
Brown booby *Sula leucogaster*	1.15	1.51	17.3
White-tailed tropicbird *Phaethon lepturus*	0.362	0.95	12.7
Double-crested cormorant *Phalacrocorax auritus*	1.42	1.16	29.8
Magnificent frigatebird *Fregata magnificens*	1.49	2.25	12.5
Great blue heron *Ardea herodias*	1.87	1.76	20.8

Table 8.4 (contd.)

White ibis			
Eudocimus albus	0.970	0.990	27.0
Turkey vulture			
Cathartes aura	2.22	1.80	23.4
Bald eagle			
Haliaeetus leucocephalus	4.20	2.18	31.1
Tundra swan			
Cygnus columbianus	9.50	2.20	64.4
Rüppell's griffon vulture			
Gyps rueppellii	7.57	2.41	45.6

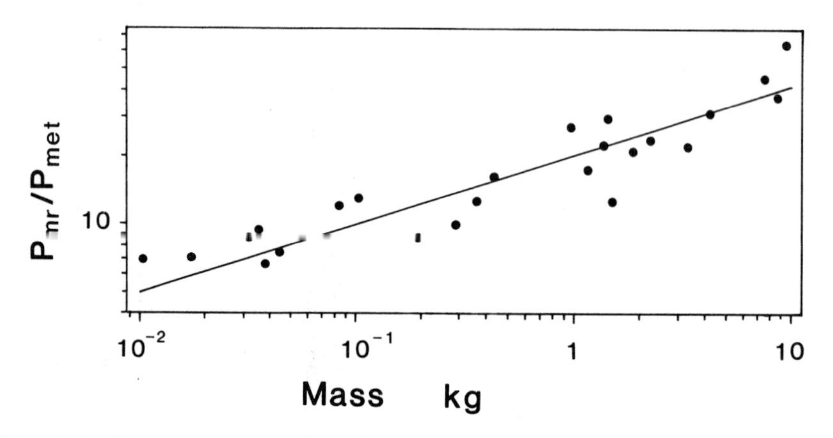

Fig. 8.6. *The power required to fly at* v_{mr} is not *a fixed multiple of the basal metabolic rate! The calculated value of this ratio increases with the 0.35 power of the mass in this sample of 24 birds of various sizes. There is also a large amount of scatter, due to some birds having longer wings than others, in relation to their mass.*

will want to compare the output with their own field observations. Such comparisons are very important for determining whether the programs give consistent estimates, and if so, whether these estimates are consistently high or low. For example, if Program 1 says that your migrating plover or duck can go only so far, when you know for certain that it goes considerably further, the implication is that the program has overestimated one or more of the calculated components of power. In a few minutes you can do several runs in which you try lowering the body drag coefficient, or the profile power ratio, or both together, to see how far you have to go before the predicted range tallies with your observation. Example 4 illustrates this kind of test.

Amending the default values

No amount of observation will produce consistent results if the underlying theory is fundamentally wrong, but this seems unlikely in view of its well-tested aeronautical origins. On the other hand, Chapter 5 should convince the reader that some of the default values for variables used in the programs are based on empirical evidence that can only be described as insecure. Some of these assumed values could be a long way out. If the underlying theory is realistic, then an accumulation of field and laboratory observations will allow these numbers to be estimated with progressively increasing confidence. Any such improvement will be based on observations of particular species, but this does not mean, as in physiology, that the results only apply to those species. This theory is supposed to be general, which means that improved accuracy of prediction will apply to species other than those actually observed.

The default values are installed at the beginning of each program, in the block of statements beginning at Line 200 (see Appendix 1). These statements can easily be edited to change the default values. The user should be wary of making such changes, however, and should only do so in response to a solid accumulation of empirical evidence. Publications that use the output of these programs should state what values were used for variables, if different from the original defaults, so that other users can reproduce the results exactly.

Appendix 1. Program listings and specimen output

The program listings in this appendix are in Roman type. Headings, and other information not constituting part of the programs, are in italics. The reader is advised to read Chapter 4 before entering Programs 1 or 1A, and Chapters 4 and 6 before entering Program 2. Nine tables of specimen output, for testing the programs, follow the listings.

The programs are copyright by C. J. Pennycuick (1988), but there are no restrictions on copying them. Results obtained with them should be acknowledged as indicated at the beginning of the output

Program 1

```
1 REM Please read Chapter 4 before typing in program
10 REM Program 1 - power curve including metabolic
components
20 REM Supplies fat consumption for a forager with variable
crop load, or range of a long-distance migrant
30 REM Body drag from Maryland results, based on all-up mass
40 REM Metabolic power based on empty body mass
50 dd=3' Set number of decimal digits for rounding output
100 DEF FNr(X)=INT((X*(10^(dd-1-INT(LOG(X)/LOG(10)))))+.5)/
(10^(dd-1-INT(LOG(X)/LOG(10))))
110 LPRINT "Quote Program 1 from: Bird Flight Performance. A
Practical Calculation"
120 LPRINT "Manual, by C.J. Pennycuick. Oxford University
Press, 1989."
150 INPUT "Species name";sp$
160 INPUT "Empty body mass (kg)";mempty
170 INPUT "Payload mass (kg)";mcrop
180 INPUT "Wing span (m)";b
200 e=3.9E+07'    Fuel energy density (J/kg)
210 eta=.23'      Conversion efficiency
220 k=1.2'        Induced power factor
230 r=1.1'        Circulation/respiration factor
240 g=9.81'       Acceleration due to gravity (m s-2)
250 rho=1.23'     Sea level air density (kg m-3)
260 x1=1.2'       Profile power ratio
270 taxon=2'        1 for passerine, 2 for non-passerine
280 hw=0'         Headwind zero
290 mradio=0:Aradio=0' No radio
330 m=mempty+mcrop
340 REM Find body frontal area (Sbody)
350 sbody=.00813*(m^.666)
360 REM Guess body Reynolds number
370 rb=125000!*SQR(m)
380 REM Find body drag coefficient
390 IF rb<50000! THEN Cdb=.4:GOTO 500
400 IF rb>200000! THEN Cdb=.25:GOTO 500
410 Cdb=1.57-(.108*LOG(rb))
500 PRINT:PRINT "Change variables or run. USE INDICATED
UNITS."
510 PRINT "1.  Species";TAB(35);sp$
520 PRINT "2.  Empty mass";TAB(35);mempty;TAB(45);"kg"
530 PRINT "3.  Payload mass";TAB(35);mcrop;TAB(45);"kg"
540 PRINT "4.  Wing span";TAB(35);b;TAB(45);"m"
```

```
550 PRINT "5.  Fuel energy density";TAB(35);e;TAB(45);"J/kg"
560 PRINT "6.  Conversion efficiency";TAB(35);eta
570 PRINT "7.  Induced power factor";TAB(35);k
580 PRINT "8.  Circulation/resp factor";TAB(35);r
590 PRINT "9.  Body frontal area";TAB(35);FNr(sbody);
TAB(45);"sq m"
595 PRINT "    and body Cd";TAB(35);FNr(Cdb)
600 PRINT "10. Air density";TAB(35);FNr(rho);TAB(45);"kg
m-3"
610 PRINT "11. Profile power ratio";TAB(35);x1
620 PRINT "12. Type of bird: ";TAB(36);
630 IF taxon=1 THEN PRINT "passerine" ELSE PRINT
"non-passerine"
640 PRINT "13. Headwind";TAB(35);
650 IF hw=0 THEN PRINT " zero":GOTO 680
660 PRINT hw;"m/s";
670 IF hw<0 THEN PRINT " (tailwind negative)" ELSE PRINT
680 PRINT "14. Run standard program"
690 PRINT "15. Find range for long-distance migrant"
694 IF mradio=0 THEN PRINT "16. Add radio":PRINT:GOTO 700
696 PRINT "16. Radio mass";TAB(35);FNr(mradio);"kg"
697 PRINT "    Radio flat-plate area";TAB(35);Aradio;"sq m"
700 INPUT "Select option  ",sel
710 ON sel GOTO
3000,3100,4000,3200,3300,3400,3500,3600,3700,3800,3900,4100,
4200,720,6000,4300
720 PRINT:INPUT "Version",sex$
730 A=(sbody*Cdb)+Aradio    ´Equivalent flat-plate area
740 m=mempty+mcrop+mradio ´Revise mass to include radio
750 Sd=3.142*(b^2)/4       ´Disc area
760 REM Find absolute minimum power (Pam)
770 Pam=.877*(k^.75)*((m*g)^1.5)*(A^.25)/(SQR(rho)*(Sd^.75))
790 Ppro=x1*Pam       ´Profile power
800 REM Find BMR and multiply by eta for mechanical
equivalent
810 IF taxon=1 THEN Pmet=eta*6.25*(mempty^.724) ELSE
Pmet=eta*3.79*(mempty^.723)
820 x2=Pmet/Pam
830 GOSUB 2000
840 REM Get preliminary estimate of Vmp
850 vmp=(.76*(k^.25)*SQR(m*g))/(SQR(rho)*((Sd*A)^.25))
860 REM Select lower and upper limits of speed for power
curve
870 V1=INT(.8*vmp)  ´Lower limit
880 IF (x1+x2)<2.5 THEN V2=FNr(1.9*vmp) ELSE V2=FNr(2.3*vmp)
885 REM PROGRAM NOT RECOMMENDED OUTSIDE SPEED RANGE V1 - V2
890 vmp=0:Vmr=0:Pmin=0:Pmr=0
900 PRINT:PRINT "Working out power curve ..."

910 FOR v=V1 TO V2 STEP .1
920    REM Find parasite drag (Dpar) and power (Ppar)
930    Dpar=rho*(v^2)*A/2
940    Ppar=Dpar*v
950    REM Find induced power (Pind)
960    REM Find resultant (rf) of weight and body drag
970    rf=SQR(((m*g)^2)+(Dpar^2))
980    Pind=k*(rf^2)/(2*rho*v*Sd)
990    p=r*(Ppar+Pind+Ppro+Pmet)
```

```
1000   IF (10*v) MOD 10<>0 THEN GOTO 1020
1010   LPRINT FNr(v);TAB(14);FNr(p);TAB(29);FNr(m*g*v/p);
TAB(44);FNr(p/eta);
1015   IF v-hw>.5 THEN LPRINT
TAB(59);FNr(1000000!*p/((v-hw)*e*eta)) ELSE LPRINT
1020   IF v=V1 THEN GOTO 1050
1030   IF vmp=0 AND NOT p<p0 THEN vmp=v:Pmin=p
1040   IF Vmr=0 AND NOT v/p>n0 THEN Vmr=v:Pmr=p
1050   p0=p:n0=v/p
1060 NEXT v

1100 REM Find max lift/drag ratio
1110 Nmax=m*g*Vmr/Pmr
1120 REM Find fat consumed per unit air distance flown
1130 fatcons=(m*g)/(Nmax*e*eta)   'Kg fat per metre

1200 PRINT
1210 PRINT "Minimum power speed    =";TAB(28);FNr(vmp);"m/s"
1220 PRINT "Maximum range speed    =";TAB(28);FNr(Vmr);"m/s"
1230 PRINT "Minimum power          =";TAB(28);FNr(Pmin);"W"
1240 PRINT "Maximum range power    =";TAB(28);FNr(Pmr);"W"
1250 PRINT "Maximum effective L/D  =";TAB(28);FNr(Nmax)
1260 PRINT "Fat consumption at Vmr =";TAB(28);
FNr(fatcons*1000000!);"g/km (air distance)"
1270 PRINT
1280 GOSUB 2400
1290 INPUT "Do more (Y/N)";YN$
1300 IF YN$="Y" OR YN$="y" THEN GOTO 500

1400 END

2000 LPRINT
2010 LPRINT sp$;"   ";sex$:LPRINT
2030 LPRINT "Assumed values:":LPRINT
2040 LPRINT "Empty body mass =";mempty;"kg"
2050 LPRINT "Payload mass    =";mcrop;"kg"
2060 LPRINT "All-up mass     =";m;"kg"
2070 LPRINT "Span            =";b;"m"
2080 LPRINT "g    =";g;"m s-2";TAB(40);"(Gravity)"
2090 LPRINT "rho  =";FNr(rho);"kg m-3";TAB(40);"(Air
density)"
2100 LPRINT "e    =";e;"J/kg";TAB(40);"(Energy density of
fat)"
2110 LPRINT "eta  =";eta;TAB(40);"(Conversion efficiency)"
2120 LPRINT "k    =";k;TAB(40);"(Induced power factor)"
2130 LPRINT "r    =";r;TAB(40);"(Circulation/respiration
factor)"
2140 LPRINT "X1   =";x1;TAB(40);"(Profile power ratio)"
2150 LPRINT "X2   =";FNr(x2);TAB(40);"(Metabolic power
ratio)"
2160 LPRINT "Disk area                 =";FNr(Sd);"sq m"
2170 LPRINT "Body frontal area         =";FNr(sbody);"sq m"
2180 LPRINT "Body drag coefficient     =";FNr(Cdb)
2190 IF mradio<>0 THEN LPRINT "Radio mass                 =";
FNr(mradio);"kg"
2200 IF mradio<>0 THEN LPRINT "Radio flat-plate area   =";
Aradio;"sq m"
2210 LPRINT "Flat-plate area           =";FNr(A);"sq m"
```

```
2230 LPRINT "Pmet                          =";FNr(Pmet);"W
(mechanical)"
2240 LPRINT "BMR                           =";FNr(Pmet/eta);"W
(chemical)"
2250 LPRINT "Fat consumption for ";
2260 IF hw=0 THEN LPRINT "zero wind":GOTO 2300
2270 LPRINT ABS(hw);"m/s";
2280 IF hw>0 THEN LPRINT " headwind" ELSE LPRINT " tailwind"
2300 LPRINT
2310 LPRINT "Airspeed";TAB(15);"Power";TAB(30);"Eff. L/D";
TAB(45);"Power";TAB(60);"Fat consumption"
2320 LPRINT "m/s";TAB(15);"W mechanical";TAB(45);"W
chemical";TAB(60);"g/km (ground)"
2330 LPRINT
2399 RETURN

2400 LPRINT:LPRINT "Computed values:":LPRINT
2410 LPRINT "Vmp                           =";FNr(vmp);"m/s"
2420 LPRINT "Vmr                           =";FNr(Vmr);"m/s"
2430 LPRINT "Pmin                          =";FNr(Pmin);"W
(mechanical)"
2440 LPRINT "Pmr                           =";FNr(Pmr);"W
(mechanical)"
2450 LPRINT "L/D max                       =";FNr(Nmax)
2460 LPRINT "Fat consumption at Vmr  =";
FNr(fatcons*1000000!);"g/km (air distance)"
2470 LPRINT
2480 LPRINT "Power estimates include basal metabolism, and"
2490 LPRINT "requirements for respiration and circulation"
2599 LPRINT CHR$(12):RETURN  'Form feed for Epson printers

3000 INPUT "New species name";sp$
3010 INPUT "Empty body mass (kg)";mempty
3020 INPUT "Payload (kg)";mcrop
3030 INPUT "Wing span (m)";b
3040 GOTO 330
3100 INPUT "New empty body mass (kg)";mempty
3110 GOTO 330
3200 INPUT "New wing span (m)";b
3210 GOTO 500
3300 INPUT "New fuel energy density (J/kg)";e
3310 GOTO 500
3400 INPUT "New conversion efficiency";eta
3410 GOTO 500
3500 INPUT "New induced drag factor";k
3510 GOTO 500
3600 INPUT "New respiration/circulation factor";r
3610 GOTO 500
3700 INPUT "Change body frontal area";YN$
3710 IF YN$="N" OR YN$="n" THEN GOTO 3730
3720 INPUT "New frontal area (sq m)";sbody
3730 INPUT "Change body drag coefficient";YN$
3740 IF YN$="N" OR YN$="n" THEN GOTO 500
3750 INPUT "New body drag coefficient";Cdb
3760 GOTO 500
3800 PRINT:PRINT "Enter air density in kg/cu m, or enter 0
to get"
3810 PRINT "temperature/pressure input"
```

```
3820 PRINT
3830 INPUT "New air density (kg m-3)";rho
3840 IF rho<>0 THEN GOTO 500 ELSE GOTO 5000
3900 INPUT "New profile power ratio";x1
3910 GOTO 500
4000 INPUT "New payload mass (kg)";mcrop
4010 GOTO 330
4100 INPUT "Type 1 for passerine, 2 for non-passerine";taxon
4110 GOTO 500
4200 PRINT "Enter headwind in m/s (negative for tailwind)"
4210 INPUT "Headwind speed (m/s)";hw
4220 GOTO 500
4300 INPUT "Radio mass in kg";mradio
4310 PRINT "Radio equivalent flat-plate area can be zero if
radio fully"
4320 PRINT "covered by feathers, up to half actual frontal
area if"
4330 PRINT "radio exposed on bird's back"
4340 INPUT "Radio equivalent flat-plate area in square
metres";Aradio
4350 GOTO 500
5000 PRINT:PRINT "Enter ambient air temperature and
pressure"
5010 PRINT "Read barometer at flight location"
5020 PRINT "Do NOT correct barometric pressure to sea level"
5030 PRINT
5040 INPUT "Air temperature in degrees C";temp
5050 PRINT
5060 PRINT "Choose pressure units:  1. Pascals"
5070 PRINT "                        2. Millibars"
5080 PRINT "                        3. Inches of mercury"
5090 PRINT
5100 INPUT "Choice";sel
5110 ON sel GOTO 5200,5300,5400
5200 INPUT "Ambient barometric pressure (pascals)";pres
5210 rho=.003486*pres/(temp+273)
5220 GOTO 500
5300 INPUT "Ambient barometric pressure (millibars)";pres
5310 rho=.3486*pres/(temp+273)
5320 GOTO 500
5400 INPUT "Ambient barometric pressure (inches of
mercury)";pres
5410 rho=11.8*pres/(temp+273)
5420 GOTO 500

6000 PRINT
6010 PRINT "Find range for a long-distance migrant"
6020 PRINT
6030 PRINT "Run standard program twice, first for empty
mass, then"
6040 PRINT "with consumable fat as payload. Use air density
for"
6050 PRINT "migration altitude. Take mean of the two
estimates of"
6060 PRINT "max effective L/D, for entry to this program"
6070 PRINT
6080 INPUT "Return to main menu (Y/N)";YN$
6090 IF YN$="Y" OR YN$="y" THEN GOTO 500
```

```
6100 LPRINT "Find range for a long-distance migrant":LPRINT
6110 LPRINT sp$:LPRINT
6120 PRINT
6130 INPUT "Mean L/D max";Nav
6140 INPUT "Fat load (kg)";m1
6150 PRINT:m0=mempty
6160 PRINT "Change variables or run. USE INDICATED UNITS"
6170 PRINT
6180 PRINT "1. Empty mass            =";m0;"kg"
6190 PRINT "2. Payload (fat)         =";m1;"kg"
6200 PRINT "3. Mean effective L/D  =";FNr(Nav)
6210 PRINT "4. Find migration range"
6220 PRINT "5. Return to main menu"
6230 PRINT "6. Exit"
6240 PRINT
6250 INPUT "Select option  ",sel
6260 ON sel GOTO 7000,7100,7200,6300,6800,6600
6300 REM Find migration range
6310 m=m0+m1+mradio:f=m1/m  'Fat fraction
6320 y=e*eta*Nav*LOG(1/(1-f))/g  'Range in metres
6330 PRINT
6340 PRINT "Range =";FNr(y/1000);"km"
6400 LPRINT
6410 LPRINT "Empty mass                 =";m0;"kg"
6420 LPRINT "Fat load at takeoff      =";m1;"kg"
6430 IF mradio<>0 THEN LPRINT "Radio mass
=";mradio;"kg"
6440 LPRINT "All-up mass at takeoff =";m;"kg"
6450 LPRINT "Fat fraction               =";FNr(f)
6460 LPRINT "Mean L/D max               =";FNr(Nav)
6470 LPRINT "Range                      =";FNr(y/1000);"km (not
corrected for wind)"
6499 GOTO 6150

6600 LPRINT CHR$(12)
6610 END

6800 LPRINT CHR$(12)
6810 GOTO 500

7000 INPUT "New empty mass (kg)";m0
7010 GOTO 6150
7100 INPUT "New fat load at takeoff (kg)";m1
7110 GOTO 6150
7200 INPUT "New mean L/D max";Nav
7210 GOTO 6150
```

Program 1A

Readers who already have a working copy of Program 1 may find it easier to create Program 1A by editing Program 1, rather than typing it in directly from the listing. A list of amendments for converting Program 1 to Program 1A follows the listing of Program 1A.

```
1 REM Please read Chapter 4 before typing in program
10 REM Program 1A
20 REM Power curve with mechanical components only
30 REM Body drag from Maryland results, based on all-up mass
50 dd=3' Set number of decimal digits for rounding output
100 DEF FNr(X)=INT((X*(10^(dd-1-INT(LOG(X)/LOG(10)))))+.5)/
(10^(dd-1-INT(LOG(X)/LOG(10))))
110 LPRINT "Quote: Program 1A from Bird Flight Performance.
A Practical Calculation"
120 LPRINT "Manual, by C.J. Pennycuick. Oxford University
Press, 1989."
140 LPRINT
150 INPUT "Species name";sp$
160 INPUT "Empty body mass (kg)";mempty
170 INPUT "Payload mass (kg)";mcrop
180 INPUT "Wing span (m)";b
220 k=1.2'          Induced power factor
240 g=9.81'         Acceleration due to gravity (m s-2)
250 rho=1.23'       Sea level air density (kg m-3)
260 x1=1.2'         Profile power ratio
290 mradio=0:Aradio=0' No radio
330 m=mempty+mcrop
340 REM Find body frontal area (Sbody)
350 sbody=.00813*(m^.666)
360 REM Guess body Reynolds number
370 rb=125000!*SQR(m)
380 REM Find body drag coefficient
390 IF rb<50000! THEN Cdb=.4:GOTO 500
400 IF rb>200000! THEN Cdb=.25:GOTO 500
410 Cdb=1.57-(.108*LOG(rb))
500 PRINT:PRINT "Change variables or run. USE INDICATED
UNITS.":PRINT
510 PRINT "1.  Species";TAB(35);sp$
520 PRINT "2.  Empty mass";TAB(35);mempty;TAB(45);"kg"
530 PRINT "3.  Payload mass";TAB(35);mcrop;TAB(45);"kg"
540 PRINT "4.  Wing span";TAB(35);b;TAB(45);"m"
550 PRINT "5.  Gravity";TAB(35);g;TAB(45);"m s-2"
570 PRINT "6.  Induced drag factor";TAB(35);k
580 PRINT "7.  Body frontal area";TAB(35);FNr(sbody);
TAB(45);"sq m"
590 PRINT "8.  Body drag coefficient";TAB(35);FNr(Cdb)
600 PRINT "9.  Air density";TAB(35);FNr(rho);TAB(45);"kg
m-3"
610 PRINT "10. Profile power ratio";TAB(35);x1
680 PRINT "11. Run standard program"
685 IF mradio=0 THEN PRINT "12. Add radio":GOTO 700
690 PRINT "12. Radio mass";TAB(35);FNr(mradio);"kg"
695 PRINT "    Radio flat-plate area";TAB(35);Aradio;"sq m"
700 PRINT:INPUT "Select option  ",sel
```

```
710 ON sel GOTO
3000,3100,4000,3200,3300,3500,3600,3700,3800,3900,720,4300
720 PRINT: INPUT "Version";sex$
730 A=(sbody*Cdb)+Aradio    'Equivalent flat-plate area
740 m=mempty+mcrop+mradio 'Revise mass to include radio
750 Sd=3.142*(b^2)/4       'Disc area
760 REM Find absolute minimum power (Pam)
770 Pam=.877*(k^.75)*((m*g)^1.5)*(A^.25)/(SQR(rho)*(Sd^.75))
790 Ppro=x1*Pam           'Profile power
830 GOSUB 2000
840 REM Get preliminary estimate of Vmp
850 vmp=(.76*(k^.25)*SQR(m*g))/(SQR(rho)*((Sd*A)^.25))
860 REM Select lower and upper limits of speed for power
curve
870 V1=INT(.8*vmp)  'Lower limit
880 IF (x1+x2)<2.5 THEN V2=FNr(1.9*vmp) ELSE V2=FNr(2.3*vmp)
885 REM PROGRAM NOT RECOMMENDED OUTSIDE SPEED RANGE V1 - V2
890 vmp=0:Vmr=0:Pmin=0:Pmr=0
900 PRINT:PRINT "Working out power curve ..."

910 FOR v=V1 TO V2 STEP .1
920    REM Find parasite drag (Dpar) and power (Ppar)
930    Dpar=rho*(v^2)*A/2
940    Ppar=Dpar*v
950    REM Find induced power (Pind)
960    REM Find resultant (rf) of weight and body drag
970    rf=SQR(((m*g)^2)+(Dpar^2))
980    Pind=k*(rf^2)/(2*rho*v*Sd)
990    p=Ppar+Pind+Ppro
1000   IF (10*v) MOD 10<>0 THEN GOTO 1020
1010   LPRINT FNr(v);TAB(14);FNr(p);TAB(29);FNr(m*g*v/p)
1020   IF v=V1 THEN GOTO 1050
1030   IF vmp=0 AND NOT p<p0 THEN vmp=v:Pmin=p
1040   IF Vmr=0 AND NOT v/p>n0 THEN Vmr=v:Pmr=p
1050   p0=p:n0=v/p
1060 NEXT v

1100 REM Find max lift/drag ratio
1110 Nmax=m*g*Vmr/Pmr

1200 PRINT
1210 PRINT "Minimum power speed    =";TAB(28);FNr(vmp);"m/s"
1220 PRINT "Maximum range speed    =";TAB(28);FNr(Vmr);"m/s"
1230 PRINT "Minimum power          =";TAB(28);FNr(Pmin);"W"
1240 PRINT "Maximum range power    =";TAB(28);FNr(Pmr);"W"
1250 PRINT "Maximum effective L/D  =";TAB(28);FNr(Nmax)
1270 PRINT
1280 GOSUB 2400
1290 INPUT "Do more (Y/N)";YN$
1300 IF YN$="Y" OR YN$="y" THEN GOTO 500

1400 END

2000 LPRINT
2010 LPRINT sp$;"   ";sex$:LPRINT
2030 LPRINT "Assumed values:":LPRINT
2040 LPRINT "Empty body mass =";mempty;"kg"
2050 LPRINT "Payload mass    =";mcrop;"kg"
```

```
2060 LPRINT "All-up mass     =";m;"kg"
2070 LPRINT "Span            =";b;"m"
2080 LPRINT "g    =";g;"m s-2";TAB(40);"(Gravity)"
2090 LPRINT "rho  =";FNr(rho);"kg m-3";TAB(40);"(Air
density)"
2120 LPRINT "k    =";k;TAB(40);"(Induced power factor)"
2140 LPRINT "X1   =";x1;TAB(40);"(Profile power ratio)"
2160 LPRINT "Disk area                =";FNr(Sd);"sq m"
2170 LPRINT "Body frontal area        =";FNr(sbody);"sq m"
2180 LPRINT "Body drag coefficient    =";FNr(Cdb)
2190 IF mradio<>0 THEN LPRINT "Radio mass                =";
FNr(mradio);"kg"
2200 IF mradio<>0 THEN LPRINT "Radio flat-plate area    =";
Aradio;"sq m"
2210 LPRINT "Flat-plate area          =";FNr(A);"sq m"
2300 LPRINT
2310 LPRINT "Airspeed";TAB(15);"Power";TAB(30);"Eff. L/D"
2320 LPRINT "m/s";TAB(15);"W mechanical"
2330 LPRINT
2399 RETURN

2400 LPRINT:LPRINT "Computed values:":LPRINT
2410 LPRINT "Vmp                      =";FNr(vmp);"m/s"
2420 LPRINT "Vmr                      =";FNr(Vmr);"m/s"
2430 LPRINT "Pmin                     =";FNr(Pmin);"W
(mechanical)"
2440 LPRINT "Pmr                      =";FNr(Pmr);"W
(mechanical)"
2450 LPRINT "L/D max                  =";FNr(Nmax)
2470 LPRINT
2480 LPRINT "Power estimates include mechanical components
only"
2599 LPRINT CHR$(12):RETURN  'Form feed for Epson printers

3000 INPUT "New species name";sp$
3010 INPUT "Empty body mass (kg)";mempty
3020 INPUT "Payload (kg)";mcrop
3030 INPUT "Wing span (m)";b
3040 GOTO 330
3100 INPUT "New empty body mass (kg)";mempty
3110 GOTO 330
3200 INPUT "New wing span (m)";b
3210 GOTO 500
3300 INPUT "New acceleration due to gravity (m s-2)";g
3310 GOTO 500
3500 INPUT "New induced drag factor";k
3510 GOTO 500
3600 INPUT "New body frontal area (sq m)";sbody
3610 GOTO 500
3700 INPUT "New body drag coefficient";Cdb
3710 GOTO 500
3800 PRINT:PRINT "Enter air density in kg/cu m, or enter 0
to get"
3810 PRINT "temperature/pressure input"
3820 PRINT
3830 INPUT "New air density (kg m-3)";rho
3840 IF rho<>0 THEN GOTO 500 ELSE GOTO 5000
3900 INPUT "New profile power ratio";x1
```

```
3910 GOTO 500
4000 INPUT "New payload mass (kg)";mcrop
4010 GOTO 330
4300 INPUT "Radio mass in kg";mradio
4310 PRINT "Radio equivalent flat-plate area can be zero if
radio fully"
4320 PRINT "covered by feathers, up to half actual frontal
area if"
4330 PRINT "radio exposed on bird's back"
4340 INPUT "Radio equivalent flat-plate area in square
metres";Aradio
4350 GOTO 500
5000 PRINT:PRINT "Enter ambient air temperature and
pressure"
5010 PRINT "Read barometer at flight location"
5020 PRINT "Do NOT correct barometric pressure to sea level"
5030 PRINT
5040 INPUT "Air temperature in degrees C";temp
5050 PRINT
5060 PRINT "Choose pressure units:  1. Pascals"
5070 PRINT "                        2. Millibars"
5080 PRINT "                        3. Inches of mercury"
5090 PRINT
5100 INPUT "Choice";sel
5110 ON sel GOTO 5200,5300,5400
5200 INPUT "Ambient barometric pressure (pascals)";pres
5210 rho=.003486*pres/(temp+273)
5220 GOTO 500
5300 INPUT "Ambient barometric pressure (millibars)";pres
5310 rho=.3486*pres/(temp+273)
5320 GOTO 500
5400 INPUT "Ambient barometric pressure (inches of
mercury)";pres
5410 rho=11.8*pres/(temp+273)
5420 GOTO 500
```

Editing Program 1 to create Program 1A

Type in the following lines in place of existing lines with the same number. Delete lines altogether as indicated in italics.

```
10 REM Program 1A
20 REM Power curve with mechanical components only
```
Delete Line 40
```
110 LPRINT "Quote: Program 1A from Bird Flight Performance.
A Practical Calculation"
```
Delete Lines 200-210
Delete Line 230
Delete Lines 270-280
Delete Lines 550-710, and substitute new Lines 550-710:
```
550 PRINT "5.  Gravity";TAB(35);g;TAB(45);"m s-2"
570 PRINT "6.  Induced drag factor";TAB(35);k
580 PRINT "7.  Body frontal area";TAB(35);FNr(sbody);
TAB(45);"sq m"
590 PRINT "8.  Body drag coefficient";TAB(35);FNr(Cd)
600 PRINT "9.  Air density";TAB(35);FNr(rho);TAB(45);"kg
m-3"
610 PRINT "10. Profile power ratio";TAB(35);x1
680 PRINT "11. Run standard program"
685 IF mradio=0 THEN PRINT "12. Add radio":GOTO 700
690 PRINT "12. Radio mass";TAB(35);FNr(mradio);"kg"
695 PRINT "    Radio flat-plate area";TAB(35);Aradio;"sq m"
700 PRINT:INPUT "Select option  ",sel
710 ON sel GOTO
3000,3100,4000,3200,3300,3500,3600,3700,3800,3900,720,4300
```
Delete Lines 800-820
```
990    p=Ppar+Pind+Ppro
1010   LPRINT FNr(v);TAB(14);FNr(p);TAB(29);FNr(m*g*v/p)
```
Delete Line 1015
Delete Lines 1120-1130
Delete Line 1260
Delete Lines 2100-2110
Delete Line 2130
Delete Line 2150
Delete Lines 2230-2280
```
2310 LPRINT "Airspeed";TAB(15);"Power";TAB(30);"Eff. L/D"
2320 LPRINT "m/s";TAB(15);"W mechanical"
```
Delete Line 2460
```
2480 LPRINT "Power estimates include mechanical components
only"
```
Delete Line 2490
Delete Lines 3300-3410
```
3600 INPUT "New body frontal area (sq m)";Sbody
3610 GOTO 500
3700 INPUT "New body drag coefficient";Cd
3710 GOTO 500
```
Delete Lines 3720-3760
Delete Lines 4100-4220
Delete Lines 6000-7210

Program 2.

Although retaining some elements in common with Programs 1 and 1A, Program 2 goes off on a different track at quite an early stage. Readers are recommended to type in Program 2 directly from the listing, and not to try to create it by editing Program 1 or 1A.

```
1 REM Please read Chapter 4 before typing in program
20 REM Program 2 - Glide superpolar and cross country
performance in thermals
30 REM Body drag from Maryland results, based on all-up mass
40 REM Fuel consumption based on twice BMR
50 dd=3' Set number of decimal digits for rounding output
90 DIM table(2,11)
100 DEF FNr(X)=INT((X*(10^(dd-1-INT(LOG(X)/LOG(10)))))+.5)/
(10^(dd-1-INT(LOG(X)/LOG(10))))
110 LPRINT "Quote: Program 2 from Bird Flight Performance. A
Practical Calculation"
120 LPRINT "Manual, by C.J. Pennycuick. Oxford University
Press, 1989.":LPRINT
130 LPRINT "Glide superpolar - Bird reduces span by 'Span
Factor' at high speeds"
140 LPRINT "to get minimum sink at each speed"
150 INPUT "Species name";sp$
160 INPUT "Empty body mass (kg)";mempty
170 INPUT "Payload mass (kg)";mcrop
180 INPUT "Wing span (m)";b
190 INPUT "Wing area (sq m)";S
200 e=3.9E+07'    Fuel energy density (J/kg)
220 k=1.1'        Induced drag factor
230 delta=1'      Slope of area vs wing span line
240 g=9.81'       Acceleration due to gravity (m s-2)
250 rho=1.23'     Sea level air density (kg m-3)
260 Cdw=.014'     Profile drag coeff for wing
270 taxon=2'      1 for passerine, 2 for non-passerine
280 hw=0'         Headwind zero
290 mradio=0:Aradio=0' No radio
300 pi=4*ATN(1)
330 m=mempty+mcrop
340 REM Find body frontal area (Sbody)
350 sbody=.00813*(m^.666)
360 REM Guess body Reynolds number
370 rb=125000!*SQR(m)
380 REM Find body drag coefficient
390 IF rb<50000! THEN Cdb=.4:GOTO 500
400 IF rb>200000! THEN Cdb=.25:GOTO 500
410 Cdb=1.57-(.108*LOG(rb))

500 PRINT:PRINT "Change variables or run. USE INDICATED
UNITS."
510 PRINT "1.  Species";TAB(35);sp$
520 PRINT "2.  Empty mass";TAB(35);mempty;TAB(45);"kg"
530 PRINT "3.  Payload mass";TAB(35);mcrop;TAB(45);"kg"
540 PRINT "4.  Wing span";TAB(35);b;TAB(45);"m"
550 PRINT "5.  Wing area";TAB(35);S;TAB(45);"sq m"
560 PRINT "6.  Slope of area:span line";TAB(35);delta
570 PRINT "7.  Induced drag factor";TAB(35);k
```

```
580 PRINT "8.  Wing profile drag coeff";TAB(35);Cdw
590 PRINT "9.  Body frontal area";TAB(35);FNr(sbody);
TAB(45);"sq m"
600 PRINT "    and drag coefficient";TAB(35);FNr(Cdb)
610 PRINT "10. Gravity";TAB(35);g;TAB(45);"m s-2"
620 PRINT "11. Air density";TAB(35);FNr(rho);TAB(45);"kg
m-3"
630 PRINT "12. Fuel energy density";TAB(35);e;TAB(45);"J/kg"
640 PRINT "13. Type of bird: ";TAB(36);
650 IF taxon=1 THEN PRINT "passerine" ELSE PRINT
"non-passerine"
660 PRINT "14. Headwind";TAB(35);
670 IF hw=0 THEN PRINT " zero":GOTO 695
680 PRINT hw;"m/s";
690 IF hw<0 THEN PRINT " (tailwind negative)" ELSE PRINT
695 IF mradio=0 THEN PRINT "15. Add radio":GOTO 710
700 PRINT "15. Radio mass";TAB(35);FNr(mradio);"kg"
705 PRINT "    Radio flat-plate area";TAB(35);Aradio;"sq m"
710 PRINT "16. Run program"
720 INPUT "Select option  ",sel
730 ON sel GOTO
3000,3100,4000,3200,3400,3600,3500,3900,5500,3700,3800,3300,
4100,4200,4300,800
800 PRINT:INPUT "Version";sex$
810 IF taxon=1 THEN BMR=6.25*(mempty^.724) ELSE
BMR=3.79*(mempty^.723)
820 A=(sbody*Cdb)+Aradio   'Equivalent flat-plate area
825 m=mempty+mcrop+mradio 'Revise mass to include radio
830 GOSUB 2000
840 REM Estimate stalling speed
850 Vs=SQR((2*m*g)/(rho*S*1.6))
860 V1=INT(Vs):v=V1
870 thresh=.5
880 Vms=0:Vbg=0:Vzmin=0:Nmax=0:Vc=0
890 PRINT:PRINT "Working out superpolar ..."

900 WHILE Vc<6.2
910   beta=((8*k*(m^2)*(g^2))/
(delta*pi*Cdw*(rho^2)*(b*2)*S*(v^4)))^(1/3)
920   IF beta>1 THEN beta=1
930   Vz=((2*k*m*g)/(pi*rho*(b^2)*(beta^2)*v))+
((rho*(v^3))*((Cdw*S*(1-(delta*(1-beta))))+A)/(2*m*g))
940   IF (10*v) MOD 10<>0 THEN GOTO 960
950   LPRINT FNr(v);TAB(19);FNr(beta);TAB(39);FNr(Vz);
TAB(59);FNr(v/Vz)
960   IF v=V1 THEN GOTO 1070
970   IF Vms=0 THEN GOTO 1050
980   Vc=(10*v*(Vz-Vzlast))-Vz
990   IF Vc<thresh THEN GOTO 1050
1000  Vx=(v*Vc)/(Vz+Vc)
1010  table(0,INT(2*Vc)-1)=v:table(1,INT(2*Vc)-1)=Vx
1020  Vx=(Vbg*Vc)/(Vzbg+Vc)
1030  table(2,INT(2*Vc)-1)=Vx
1040  thresh=thresh+.5
1050  IF Vms=0 AND NOT Vz<Vzlast THEN Vms=v-.1:Vzmin=Vzlast
1060  IF Nmax=0 AND NOT v/Vz>Nlast THEN Vbg=v-.1:Nmax=Nlast:
Vzbg=Vzlast
1070  Vzlast=Vz:Nlast=v/Vz
```

```
1080  v=v+.1
1090 WEND

1310 GOSUB 2400
1320 GOSUB 2600
1330 PRINT
1340 INPUT "Do more (Y/N)";yn$
1350 IF yn$="Y" OR yn$="y" THEN GOTO 500

1400 END

2000 LPRINT
2010 LPRINT sp$;"   ";sex$:LPRINT
2030 LPRINT "Assumed values:":LPRINT
2040 LPRINT "Empty body mass =";mempty;"kg"
2050 LPRINT "Payload mass    =";mcrop;"kg"
2060 LPRINT "All-up mass     =";m;"kg"
2070 LPRINT "Span            =";b;"m"
2080 LPRINT "Wing area       =";S;"sq m"
2090 LPRINT "g    =";g;"m s-2";TAB(40);"(Gravity)"
2100 LPRINT "rho  =";FNr(rho);"kg m-3";TAB(40);"(Air
density)"
2110 LPRINT "e    =";e;"J/kg";TAB(40);"(Energy density of
fat)"
2120 LPRINT "k    =";k;TAB(40);"(Induced drag factor)"
2130 LPRINT "delta =";delta;TAB(40);"(Slope of area vs span
line)"
2140 LPRINT "Cdw  =";Cdw;TAB(40);"(Wing profile drag coeff)"
2170 LPRINT "Body frontal area        =";FNr(sbody);"sq m"
2180 LPRINT "Body drag coefficient    =";FNr(Cdb)
2190 IF mradio<>0 THEN LPRINT "Radio mass                =";
FNr(mradio);"kg"
2200 IF mradio<>0 THEN LPRINT "Radio flat-plate area    =";
Aradio;"sq m"
2210 LPRINT "Flat-plate area          =";FNr(A);"sq m"
2230 LPRINT "BMR                      =";FNr(BMR);"W
(chemical)"
2320 LPRINT
2330 LPRINT "True airspeed";TAB(20);"Span factor";TAB(40);
"Sinking speed";TAB(60);"Glide ratio"
2340 LPRINT " m/s";TAB(40);" m/s"
2399 RETURN

2400 LPRINT:LPRINT "Computed values:":LPRINT
2410 LPRINT "Stall speed (Cl=1.6) =";TAB(25);FNr(Vs);"m/s"
2420 LPRINT "Minimum sink         =";TAB(25);
2430 IF Vms=0 THEN LPRINT "(below stall speed)" ELSE LPRINT
FNr(Vzmin);"m/s   at";TAB(42);FNr(Vms);"m/s"
2440 LPRINT "Best glide ratio     =";TAB(25);FNr(Nmax);"
at";TAB(42);FNr(Vbg);"m/s"
2450 LPRINT "Circling radius
=";TAB(25);FNr(3.51*m/(S*rho));"m at 24 degrees bank and
Cl=1.4"
2590 LPRINT CHR$(12)
2599 RETURN

2600 LPRINT sp$;"   ";sex$;"   (continued)":LPRINT
2610 LPRINT "Optimum interthermal speed (Vopt) and
```

```
cross-country speed vs rate of climb"
2615 LPRINT "in thermals"
2620 LPRINT:LPRINT "Vxc (opt) is cross-country speed if bird
flies at Vopt between thermals"
2630 LPRINT "Vxc (bg) is cross-country speed if bird flies
at Vbg between thermals"
2635 LPRINT "Vbg =";FNr(Vbg);"m/s (best glide speed)":LPRINT
2640 LPRINT "Fat consumption for ";
2650 IF hw=0 THEN LPRINT "zero wind";:GOTO 2680
2660 LPRINT ABS(hw);"m/s";
2670 IF hw>0 THEN LPRINT " headwind"; ELSE LPRINT "
tailwind";
2680 LPRINT "  (based on 2xBMR and Vxc (opt))":LPRINT
2710 LPRINT "Climb";TAB(15);"Vopt";TAB(30);"Vxc
(opt)";TAB(45);"Vxc (bg)";TAB(60);"Fat cons (ground)"
2720 LPRINT "m/s";TAB(15);"m/s";TAB(30);"m/s";TAB(45);"m/s";
TAB(60);"g/km":LPRINT
2730 FOR j=0 TO 11
2740 LPRINT (j+1)/2;TAB(14);FNr(table(0,j));TAB(29);
FNr(table(1,j));TAB(44);FNr(table(2,j));TAB(59);
2750 IF table(1,j)-hw>.5 THEN LPRINT
FNr((2000000!*BMR)/((table(1,j)-hw)*e)) ELSE LPRINT "***"
2760 NEXT j
2790 LPRINT CHR$(12)
2799 RETURN

3000 INPUT "New species name";sp$
3010 INPUT "Empty body mass (kg)";mempty
3020 INPUT "Payload (kg)";mcrop
3030 INPUT "Wing span (m)";b
3040 INPUT "Wing area (sq m)";S
3050 GOTO 330
3100 INPUT "New empty body mass (kg)";mempty
3110 GOTO 330
3200 INPUT "New wing span (m)";b
3210 GOTO 500
3300 INPUT "New fuel energy density (J/kg)";e
3310 GOTO 500
3400 INPUT "Wing area";S
3410 GOTO 500
3500 INPUT "New induced drag factor";k
3510 GOTO 500
3600 INPUT "New slope for area vs span line";delta
3610 GOTO 500
3700 INPUT "New acceleration due to gravity (m s-2)";g
3710 GOTO 500
3800 PRINT:PRINT "Enter air density in kg/cu m, or enter 0
to get"
3810 PRINT "temperature/pressure input"
3820 PRINT
3830 INPUT "New air density (kg m-3)";rho
3840 IF rho<>0 THEN GOTO 500 ELSE GOTO 5000
3900 INPUT "Wing profile drag coefficient";Cdw
3910 GOTO 500
4000 INPUT "New payload mass (kg)";mcrop
4010 GOTO 330
4100 INPUT "Type 1 for passerine, 2 for non-passerine";taxon
4110 GOTO 500
```

```
4200 PRINT "Enter headwind in m/s (negative for tailwind)"
4210 INPUT "Headwind speed (m/s)";hw
4220 GOTO 500
4300 INPUT "Radio mass in kg";mradio
4310 PRINT "Radio equivalent flat-plate area can be zero if
radio fully"
4320 PRINT "covered by feathers, up to half actual frontal
area if"
4330 PRINT "radio exposed on bird's back"
4340 INPUT "Radio equivalent flat-plate area in square
metres";Aradio
4350 GOTO 500
5000 PRINT:PRINT "Enter ambient air temperature and
pressure"
5010 PRINT "Read barometer at flight location"
5020 PRINT "Do NOT correct barometric pressure to sea level"
5030 PRINT
5040 INPUT "Air temperature in degrees C";temp
5050 PRINT
5060 PRINT "Choose pressure units:  1. Pascals"
5070 PRINT "                        2. Millibars"
5080 PRINT "                        3. Inches of mercury"
5090 PRINT
5100 INPUT "Choice";sel
5110 ON sel GOTO 5200,5300,5400
5200 INPUT "Ambient barometric pressure (pascals)";pres
5210 rho=.003486*pres/(temp+273)
5220 GOTO 500
5300 INPUT "Ambient barometric pressure (millibars)";pres
5310 rho=.3486*pres/(temp+273)
5320 GOTO 500
5400 INPUT "Ambient barometric pressure (inches of
mercury)";pres
5410 rho=11.8*pres/(temp+273)
5420 GOTO 500
5500 INPUT "Change body frontal area (Y/N)";yn$
5510 IF yn$="N" OR yn$="n" THEN GOTO 5530
5520 INPUT "New body frontal area (sq m)";sbody
5530 INPUT "Change body drag coefficient (Y/N)";yn$
5540 IF yn$="N" OR yn$="n" THEN GOTO 500
5550 INPUT "New body drag coefficient";Cdb
5560 GOTO 500
```

The tables on the following 10 pages are reproduced exactly as they were printed by the computer. They are intended as test examples to check that the programs are working correctly. Information about the examples, and instructions for running them, are in Chapter 8. Before the programs can be used for original research, it is essential that they reproduce these examples correctly. The numbers, and also the layout of the output tables, must be exactly as shown. Any discrepancies will be due to errors in typing in the program, or syntax differences between different versions of BASIC. Advice on dealing with such problems is in Chapter 4.

As no additional headings have been printed on the tables, they have to be identified by order of appearance as follows:

Table A1.1. Pigeon (light) from Rothe et al's (1987) wind tunnel experiment. Note that the air density has been changed from the default value. Output from Program 1.

Table A1.2. Pigeon (heavy), same experiment.

Table A1.3. Pigeon (heavy), with headwind.

Table A1.4. Snow goose at takeoff with full fuel. Air density 1 kg m^{-3} for cruising altitude. Program 1.

Table A1.5. Snow goose. Output from subsidiary range program of Program 1, using results from Tables A1.4, and a second run (output not shown) with zero payload.

Table A1.6. Snow goose with full fuel and early satellite trackable radio trasmitter. Program 1.

Table A1.7. Two runs of range program of Program 1, based on Table A1.6, and 3 further runs of the main program. Range with radio as in Table A1.6, and then with radio of same mass but no drag.

Table A1.8. Double-crested cormorant. Output from Program 1A.

Table A1.9. Gliding performance of the African white-backed vulture. Output from Program 2 (2 pages).

Quote Program 1 from: Bird Flight Performance. A Practical Calculation Manual, by C.J. Pennycuick. Oxford University Press, 1989.

Pigeon light

Assumed values:

```
Empty body mass = .31 kg
Payload mass    = 0 kg
All-up mass     = .31 kg
Span            = .6 m
g     = 9.81 m s-2                        (Gravity)
rho   = 1.19 kg m-3                       (Air density)
e     = 3.9E+07 J/kg                      (Energy density of fat)
eta   = .23                               (Conversion efficiency)
k     = 1.2                               (Induced power factor)
r     = 1.1                               (Circulation/respiration factor)
X1    = 1.2                               (Profile power ratio)
X2    = .154                              (Metabolic power ratio)
Disk area               = .283 sq m
Body frontal area       = .00373 sq m
Body drag coefficient   = .366
Flat-plate area         = .00136 sq m
Pmet                    = .374 W (mechanical)
BMR                     = 1.63 W (chemical)
```

Fat consumption for zero wind

Airspeed m/s	Power W mechanical	Eff. L/D	Power W chemical	Fat consumption g/km (ground)
7	6.51	3.27	28.3	.104
8	6.33	3.84	27.5	.0883
9	6.28	4.36	27.3	.0777
10	6.32	4.82	27.5	.0704
11	6.45	5.19	28	.0653
12	6.66	5.48	29	.0619
13	6.97	5.67	30.3	.0597
14	7.36	5.79	32	.0586
15	7.83	5.82	34.1	.0582
16	8.4	5.79	36.5	.0585
17	9.06	5.7	39.4	.0594

Computed values:

```
Vmp                     = 9.2 m/s
Vmr                     = 15.1 m/s
Pmin                    = 6.28 W (mechanical)
Pmr                     = 7.89 W (mechanical)
L/D max                 = 5.82
Fat consumption at Vmr  = .0582 g/km (air distance)
```

Power estimates include basal metabolism, and requirements for respiration and circulation

geon heavy

sumed values:

pty body mass = .35 kg
yload mass = 0 kg
l-up mass = .35 kg
an = .6 m
= 9.81 m s-2 (Gravity)
ɔ = 1.19 kg m-3 (Air density)
= 3.9E+07 J/kg (Energy density of fat)
a = .23 (Conversion efficiency)
= 1.2 (Induced power factor)
= 1.1 (Circulation/respiration factor)
= 1.2 (Profile power ratio)
= .138 (Metabolic power ratio)
sk area = .283 sq m
dy frontal area = .00404 sq m
dy drag coefficient = .359
at-plate area = .00145 sq m
et = .408 W (mechanical)
R = 1.77 W (chemical)
t consumption for zero wind

rspeed s	Power W mechanical	Eff. L/D	Power W chemical	Fat consumption g/km (ground)
	7.87	3.01	34.7	.127
	7.72	3.56	33.8	.108
	7.61	4.06	33.1	.0942
0	7.61	4.51	33.1	.0848
1	7.71	4.9	33.5	.0782
2	7.92	5.2	34.4	.0735
3	8.21	5.43	35.7	.0704
4	8.61	5.58	37.4	.0685
5	9.1	5.66	39.6	.0676
6	9.69	5.67	42.1	.0675
7	10.4	5.62	45.1	.0681

mputed values:

p = 9.6 m/s
r = 15.8 m/s
in = 7.6 W (mechanical)
r = 9.56 W (mechanical)
D max = 5.67
t consumption at Vmr = .0675 g/km (air distance)

wer estimates include basal metabolism, and
quirements for respiration and circulation

Pigeon   heavy with headwind

Assumed values:

```
Empty body mass = .35 kg
Payload mass    = 0 kg
All-up mass     = .35 kg
Span            = .6 m
g    = 9.81 m s-2                        (Gravity)
rho  = 1.19 kg m-3                       (Air density)
e    = 3.9E+07 J/kg                      (Energy density of fat)
eta  = .23                               (Conversion efficiency)
k    = 1.2                               (Induced power factor)
r    = 1.1                               (Circulation/respiration factor)
X1   = 1.2                               (Profile power ratio)
X2   = .138                              (Metabolic power ratio)
Disk area               = .283 sq m
Body frontal area       = .00404 sq m
Body drag coefficient   = .359
Flat-plate area         = .00145 sq m
Pmet                    = .408 W (mechanical)
BMR                     = 1.77 W (chemical)
Fat consumption for  8 m/s headwind
```

Airspeed m/s	Power W mechanical	Eff. L/D	Power W chemical	Fat consumpti g/km (ground)
7	7.97	3.01	34.7	
0	7.72	3.56	33.6	
9	7.61	4.06	33.1	.848
10	7.61	4.51	33.1	.424
11	7.71	4.9	33.5	.287
12	7.92	5.2	34.4	.221
13	8.21	5.43	35.7	.183
14	8.61	5.58	37.4	.16
15	9.1	5.66	39.6	.145
16	9.69	5.67	42.1	.135
17	10.4	5.62	45.1	.129

Computed values:

```
Vmp                      = 9.6 m/s
Vmr                      = 15.8 m/s
Pmin                     = 7.6 W (mechanical)
Pmr                      = 9.56 W (mechanical)
L/D max                  = 5.67
Fat consumption at Vmr   = .0675 g/km (air distance)
```

Power estimates include basal metabolism, and
requirements for respiration and circulation

uote Program 1 from: Bird Flight Performance. A Practical Calculation
anual, by C.J. Pennycuick. Oxford University Press, 1989.

now goose   Full fuel no radio

ssumed values:

mpty body mass = 2.59 kg
ayload mass = .65 kg
ll-up mass = 3.24 kg
pan = 1.6 m
= 9.81 m s-2 (Gravity)
ho = 1 kg m-3 (Air density)
= 3.9E+07 J/kg (Energy density of fat)
ta = .23 (Conversion efficiency)
= 1.2 (Induced power factor)
= 1.1 (Circulation/respiration factor)
1 = 1.2 (Profile power ratio)
2 = .063 (Metabolic power ratio)
isk area = 2.01 sq m
ody frontal area = .0178 sq m
ody drag coefficient = .25
lat-plate area = .00445 sq m
met = 1.73 W (mechanical)
MR = 7.54 W (chemical)
at consumption for zero wind

irspeed /s	Power W mechanical	Eff. L/D	Power W chemical	Fat consumption g/km (ground)
11	71.7	4.88	312	.726
12	70.1	5.44	305	.652
13	69.2	5.97	301	.593
14	68.7	6.48	299	.547
15	68.6	6.95	298	.51
16	69	7.37	300	.481
17	69.8	7.74	304	.458
18	71	8.06	309	.44
19	72.5	8.33	315	.425
20	74.4	8.54	324	.415
21	76.7	8.7	334	.407
22	79.4	8.81	345	.402
23	82.5	8.86	359	.4
24	85.9	8.88	374	.399
25	89.8	8.85	390	.4
26	94	8.79	409	.403
27	98.7	8.69	429	.408

omputed values:

mp = 14.7 m/s
mr = 23.9 m/s
min = 68.6 W (mechanical)
mr = 85.6 W (mechanical)
/D max = 8.88
at consumption at Vmr = .399 g/km (air distance)

ower estimates include basal metabolism, and
equirements for respiration and circulation

```
Find range for a long-distance migrant

Snow goose


Empty mass               = 2.59 kg
Fat load at takeoff      = .65 kg
All-up mass at takeoff   = 3.24 kg
Fat fraction             = .201
Mean L/D max             = 9.18
Range                    = 1880 km (not corrected for wind)
```

```
ow goose   Full fuel radio 3A

sumed values:

pty body mass = 2.59 kg
yload mass    = .65 kg
l-up mass     = 3.4 kg
an            = 1.6 m
  = 9.81 m s-2                      (Gravity)
o = 1 kg m-3                        (Air density)
  = 3.9E+07 J/kg                    (Energy density of fat)
a = .23                             (Conversion efficiency)
  = 1.2                             (Induced power factor)
  = 1.1                             (Circulation/respiration factor)
  = 1.2                             (Profile power ratio)
  = .055                            (Metabolic power ratio)
sk area               = 2.01 sq m
dy frontal area       = .0178 sq m
dy drag coefficient   = .25
dio mass              = .16 kg
dio flat-plate area   = .00128 sq m
at-plate area         = .00573 sq m
et                    = 1.73 W (mechanical)
R                     = 7.54 W (chemical)
t consumption for zero wind
```

rspeed s	Power W mechanical	Eff. L/D	Power W chemical	Fat consumption g/km (ground)
1	80.9	4.53	352	.82
2	79.4	5.04	345	.738
3	78.6	5.52	342	.674
4	78.3	5.96	340	.623
5	78.5	6.37	341	.584
6	79.3	6.73	345	.552
7	80.5	7.04	350	.528
8	82.2	7.3	358	.509
9	84.4	7.51	367	.495
0	87	7.66	378	.485
1	90.1	7.77	392	.479
2	93.7	7.83	407	.475
3	97.8	7.84	425	.474
4	102	7.82	445	.475
5	107	7.76	467	.479
6	113	7.67	491	.485

```
mputed values:

p                     = 14.1 m/s
r                     = 23 m/s
in                    = 78.3 W (mechanical)
r                     = 97.8 W (mechanical)
D max                 = 7.84
t consumption at Vmr  = .474 g/km (air distance)

wer estimates include basal metabolism, and
quirements for respiration and circulation
```

```
Find range for a long-distance migrant

Snow goose


Empty mass               = 2.59 kg
Fat load at takeoff      = .65 kg
Radio mass               = .16 kg
All-up mass at takeoff = 3.4 kg
Fat fraction             = .191
Mean L/D max             = 8.04
Range                    = 1560 km (not corrected for wind)

Empty mass               = 2.59 kg
Fat load at takeoff      = .65 kg
Radio mass               = .16 kg
All-up mass at takeoff = 3.4 kg
Fat fraction             = .191
Mean L/D max             = 9.2
Range                    = 1780 km (not corrected for wind)
```

uote: Program 1A from Bird Flight Performance. A Practical Calculation
anual, by C.J. Pennycuick. Oxford University Press, 1989.

ouble-crested cormorant

ssumed values:

```
mpty body mass = 1.41 kg
ayload mass    = 0 kg
ll-up mass     = 1.41 kg
pan            = 1.16 m
   = 9.81 m s-2                          (Gravity)
ho = 1.23 kg m-3                         (Air density)
   = 1.2                                 (Induced power factor)
1  = 1.2                                 (Profile power ratio)
isk area               = 1.06 sq m
ody frontal area       = .0102 sq m
ody drag coefficient   = .284
lat-plate area         = .0029 sq m
```

irspeed /s	Power W mechanical	Eff. L/D
9	23.6	5.28
10	23.1	5.99
11	22.9	6.65
12	22.9	7.25
13	23.2	7.76
14	23.7	8.18
15	24.4	8.51
16	25.3	8.75
17	26.4	8.9
18	27.8	8.96
19	29.4	8.95
20	31.2	8.88
21	33.2	8.75

omputed values:

```
mp                     = 11.4 m/s
mr                     = 18.5 m/s
min                    = 22.9 W (mechanical)
mr                     = 28.5 W (mechanical)
/D max                 = 8.96
```

ower estimates include mechanical components only

Quote: Program 2 from Bird Flight Performance. A Practical Calculatio
Manual, by C.J. Pennycuick. Oxford University Press, 1989.

Glide superpolar - Bird reduces span by 'Span Factor' at high speeds
to get minimum sink at each speed

White-backed vulture

Assumed values:

```
Empty body mass = 5.38 kg
Payload mass    = 0 kg
All-up mass     = 5.38 kg
Span            = 2.18 m
Wing area       = .69 sq m
g     = 9.81 m s-2                          (Gravity)
rho   = 1.23 kg m-3                         (Air density)
e     = 3.9E+07 J/kg                        (Energy density of fat)
k     = 1.1                                 (Induced drag factor)
delta = 1                                   (Slope of area vs span line)
Cdw   = .014                                (Wing profile drag coeff)
Body frontal area       = .0249 sq m
Body drag coefficient   = .25
Flat-plate area         = .00623 sq m
BMR                     = 12.8 W (chemical)
```

True airspeed m/s	Span factor	Sinking speed m/s	Glide rati
8	1	.885	9.04
9	1	.838	10.7
10	1	.817	12.2
11	1	.821	13.4
12	1	.847	14.2
13	1	.893	14.6
14	1	.96	14.6
15	1	1.05	14.3
16	1	1.15	13.9
17	1	1.28	13.3
18	1	1.43	12.6
19	.979	1.6	11.9
20	.915	1.78	11.2
21	.857	1.98	10.6
22	.806	2.18	10.1
23	.759	2.4	9.58
24	.717	2.63	9.12
25	.679	2.88	8.69
26	.645	3.14	8.29
27	.613	3.41	7.92
28	.584	3.7	7.57
29	.557	4	7.24
30	.533	4.32	6.94
31	.51	4.66	6.65

Computed values:

```
Stall speed (Cl=1.6) =  8.82 m/s
Minimum sink         =  .816 m/s   at   10.3 m/s
Best glide ratio     =  14.6       at   13.6 m/s
Circling radius      =  22.3 m at 24 degrees bank and Cl=1.4
```

hite-backed vulture (continued)

ptimum interthermal speed (Vopt) and cross-country speed vs rate of climb n thermals

xc (opt) is cross-country speed if bird flies at Vopt between thermals
xc (bg) is cross-country speed if bird flies at Vbg between thermals
bg = 13.6 m/s (best glide speed)

at consumption for zero wind (based on 2xBMR and Vxc (opt))

limb /s	Vopt m/s	Vxc (opt) m/s	Vxc (bg) m/s	Fat cons (ground) g/km
.5	15.4	4.96	4.85	.132
1	16.9	7.54	7.12	.087
1.5	18.2	9.26	8.43	.0708
2	20.3	10.6	9.32	.0617
2.5	22.2	11.8	9.94	.0556
3	23.8	12.8	10.4	.0512
3.5	25.3	13.8	10.8	.0477
4	26.6	14.6	11	.045
4.5	27.9	15.4	11.3	.0426
5	29	16.1	11.5	.0407
5.5	30.1	16.8	11.6	.039
6	31.1	17.5	11.8	.0376

Appendix 2. Units and dimensions

An introduction to the SI system and its relationship to other commonly used unit systems may be found in Pennycuick (1988), together with an extensive selection of conversion factors. The following summary is for quick reference.

Mass. Dimensions M.
SI unit: kilogram (kg) (arbitrarily defined).

Length. Dimensions L.
SI unit: metre (m) (arbitrarily defined).

Time. Dimensions T.
SI unit: second (s) (arbitrarily defined).

Area. Dimensions L^2.
SI unit: square metre (m^2).

Volume. Dimensions L^3.
SI unit: cubic metre (m^3).

Speed. Dimensions LT^{-1}.
SI unit: metre per second ($m\ s^{-1}$).

Density. Dimensions ML^{-3}.
SI unit: kilogram per cubic metre ($kg\ m^{-3}$).

Force, weight. Dimensions MLT^{-2}.
SI unit: newton (N).
1 newton = $1\ kg\ m\ s^{-2}$.

Pressure, stress. Dimensions $ML^{-1}T^{-2}$.
SI unit: pascal (Pa).
1 pascal = $1\ N\ m^{-2}$.

Work, energy. Dimensions ML^2T^{-2}.
SI unit: joule (J).
1 joule = 1 N m.
(Note that nm is short for nanometre, NM is nautical mile.)

Power. Dimensions ML^2T^{-3}.
SI unit: watt (W).
$1\ W = 1\ J\ s^{-1}$.

Energy density. Dimensions L^2T^{-2}.
SI unit: joule per kilogram ($J\ kg^{-1}$).

Kinematic viscosity. Dimensions L^2T^{-1}.
SI unit: square metre per second (m^2s^{-1}).

Appendix 3. List of symbols

Each symbol has only one meaning within any given chapter, but a few of them have different meanings in different chapters. Duplication occurs mainly in Chapter 7, where total consistency would have meant changing some familiar usages, with little or no benefit in clarity. The page numbers indicate where each symbol is introduced and defined.

Symbol	Meaning	Page
A	Equivalent flat-plate area	20
C	Body circumference	14
C_D	Drag coefficient	45
C_{Db}	Body drag coefficient	20
C_{Dpro}	Profile drag coefficient	53
C_f	Rate of fat consumption	29
C_J	Resultant force coefficient	65
C_L	Lift coefficient	54
D	Drag	45
D_{ind}	Induced drag	67
D_{par}	Parasite drag	19
D_{pro}	Profile drag	53
F	Fat fraction	29
F	Force	82
F_0	Isometric force	84
F_m	Mass rate of flow	22
J	Resultant aerodynamic force	65
L	Length (in dimensions)	7
L	Lift	54
L	Length of muscle (extended)	82
ΔL	Shortening distance	82
M	Mass (in dimensions)	7
M	Moment	92
N	Effective lift:drag ratio	29
P	Power	17

P_F	Power required to maintain force	94
P_{aer}	Total power to fly aerobically	27
P_{an}	Total power to fly anaerobically	27
P_{am}	Absolute minimum power	24
P_{ind}	Induced power	21
P_m	Mass-specific power	83
P_{met}	Metabolic power	27
P_{par}	Parasite power	18
P_{pro}	Profile power	24
P_v	Volume-specific power	83
Q	Work	15
Q_m	Mass-specific work	83
Q_v	Volume-specific work	83
R	Respiration and circulation factor	27
Re	Reynolds number	42
S	Wing area	10
S	Cross-sectional area of muscle	82
S_b	Frontal area of body	14
S_d	Disc area	22
T	Time (in dimensions)	7
T	Thrust	19
T	Temperature	15
V	True airspeed	3
V	Speed of shortening (muscle)	84
V_{bg}	Speed for best glide ratio	66
V_c	Rate of climb	76
V_i	Induced velocity	23
V_{it}	Inter-thermal speed	76
V_g	Groundspeed	29
V_{min}	Minimum gliding speed	66
V_{mp}	Minimum power speed	3
V_{mr}	Maximum range speed	3
V_{ms}	Minimum sink speed	66
V_{opt}	Optimum inter-thermal speed	77
V_{xc}	Cross-country speed	76
V_z	Vertical component of speed	64
W	Wing loading	14
X_1	Profile power ratio	26
Y	Range	29
a	Hill's force constant	84
b	Wing span	10
b	Hill's velocity constant	84
c	Wing chord	13
e	Energy density of fuel	29
f	Frequency	83
g	Acceleration due to gravity	9

j_1	Moment arm of lift on wing	92
j_2	Moment arm of pectoralis	92
k	Induced power (or drag) factor	23
k	Inverse power density of mitochondria	89
l	Length	42
m	All-up body mass	9
p	Pressure	15
r	Turning radius	75
v	Volume of muscle	82
v_c	Volume of myofibrils	89
v_t	Volume of mitochondria	89
Λ	Aspect ratio	14
α	Glide angle	64
α	Hill's stress constant	85
β	Wing span reduction factor	70
β	Hill's strain rate constant	85
δ	Planform slope	70
ϵ	Wing area reduction factor	70
ϕ	Angle of bank	75
η	Conversion efficiency	16
λ	Active strain	82
ν	Kinematic viscosity	15
ν	Strain rate	85
ρ	Air density	14
ρ	Muscle density	83
σ	Stress	82
σ_0	Isometric stress	85

Appendix 4. Some properties of the standard atmosphere

Altitude (m above sea level)	Air density ($kg\ m^{-3}$)	Kinematic viscosity ($m^2\ s^{-1}$)
0	1.23	1.45×10^{-5}
500	1.17	1.52×10^{-5}
1000	1.11	1.58×10^{-5}
1500	1.06	1.64×10^{-5}
2000	1.01	1.70×10^{-5}
2500	0.957	1.77×10^{-5}
3000	0.910	1.85×10^{-5}
3500	0.864	1.94×10^{-5}
4000	0.820	2.02×10^{-5}
4500	0.777	2.11×10^{-5}
5000	0.736	2.20×10^{-5}
5500	0.697	2.30×10^{-5}
6000	0.660	2.40×10^{-5}
6500	0.624	2.51×10^{-5}
7000	0.590	2.63×10^{-5}

References

Abbot, I. H. and Doenhoff, A. E. von (1959). *Theory of wing sections*. New York, Dover.

Alexander, R. McN. (1968). *Animal mechanics*. Sidgwick and Jackson, London.

Baudinette, R. V. and Schmidt-Nielsen, K. (1974). Energy cost of gliding flight in herring gulls. *Nature (London)*, **248**, 83–4.

Brown, R. H. J. (1948). The flight of birds: the flapping cycle of the pigeon. *Journal of Experimental Biology*, **25**, 322–33.

Ellington, C. P. (1978). The aerodynamics of normal hovering flight: three approaches. In *Comparative physiology: water, ions and fluid mechanics* (ed. K. Schmidt-Nielsen, L. Bolis, and S. H. P. Maddrell), pp. 327–45. Cambridge University Press, Cambridge.

Ellington, C. P. (1984). The aerodynamics of hovering flight. V. A vortex theory. *Philosophical Transactions of the Royal Society of London, B*, **305**, 115–44.

Gauthier, G., Bedard, J., Huot, J., and Bedard, Y. (1984). Spring accumulation of fat by greater snow geese in two staging habitats. *Condor*, **86**, 192–9.

George, J. C. and Berger, A. J. (1966). *Avian myology*. Academic Press, New York.

Greenewalt, C. H. (1962). Dimensional relationships for flying animals. *Smithsonian Miscellaneous Collections*, **144**, 1–46.

Hill, A. V. (1938). The heat of shortening and the dynamic constants of muscle. *Proceedings of the Royal Society of London, B*, **126**, 136–95.

Hill, A. V. (1950). The dimensions of animals and their muscular dynamics. *Science Progress*, **38**, 209–30.

Johnston, I. A. (1985). Sustained force development: specializations and variation among the vertebrates. *Journal of Experimental Biology*, **115**, 239–51.

Kerlinger, P. (1989). *Flight strategies of migrating hawks*. Chicago University Press, Chicago.

Lasiewski, R. C. and Dawson, W. R. (1967). A re-examination of the relation between standard metabolic rate and body weight in birds. *Condor*, **69**, 13–23.

McMahon, T. A. (1984). *Muscles, reflexes, and locomotion*. Princeton University Press, Princeton.

Milne-Thomson, L. M. (1958). *Theoretical aerodynamics*. Macmillan (Dover edition 1973), New York.

Mises, R. von (1945). *Theory of flight*. McGraw-Hill. (Dover edition, 1959), New York.

Norberg, U. M. (1989). *Vertebrate flight*. Springer, Berlin, Heidelberg.

Obrecht, H. H., Pennycuick, C. J., and Fuller, M. R. (1988). Wind tunnel experiments to assess the effect of back-mounted radio transmitters on bird body drag. *Journal of Experimental Biology*, **135**, 265–73.

Pennycuick, C. J. (1968*a*). A wind-tunnel study of gliding flight in the pigeon *Columba livia*. *Journal of Experimental Biology*, **49**, 509–26.

Pennycuick, C. J. (1968*b*). Power requirements for horizontal flight in the pigeon *Columba livia*. *Journal of Experimental Biology*, **49**, 527–55.

Pennycuick, C. J. (1969). The mechanics of bird migration. *Ibis*, **111**, 525–56.

Pennycuick, C. J. (1971). Control of gliding angle in Rüppell's griffon vulture *Gyps rueppellii*. *Journal of Experimental Biology*, **55**, 39–46.

Pennycuick, C. J. (1972). Soaring behaviour and performance of some East African birds, observed from a motor-glider. *Ibis*, **114**, 178–218.

Pennycuick, C. J. (1975). Mechanics of flight. In *Avian Biology*, vol. 5 (ed. D. S. Farner and J. R. King), pp. 1–75. Academic Press, New York.

Pennycuick, C. J. (1983). Thermal soaring compared in three dissimilar tropical bird species, *Fregata magnificens*, *Pelecanus occidentalis*, and *Coragyps atratus*. *Journal of Experimental Biology*, **102**, 307–25.

Pennycuick, C. J. (1987). Flight of seabirds. In *Seabirds: feeding ecology and role in marine ecosystems* (ed. J. P. Croxall), pp. 43–62. Cambridge University Press, Cambridge.

Pennycuick, C. J. (1988). *Conversion factors: SI units and many others*. University of Chicago Press, Chicago.

Pennycuick, C. J. (1989). Span-ratio analysis used to estimate effective lift:drag ratio in the double-crested cormorant *Phalacrocorax auritus*, from field observations. *Journal of Experimental Biology*, **142**, 1–15.

Pennycuick, C. J. and Rezende, M. A. (1984). The specific power output of aerobic muscle, related to the power density of mitochondria. *Journal of Experimental Biology*, **108**, 377–92.

Pennycuick, C. J., Alerstam, T., and Larsson, B. (1979). Soaring migration of the common crane *Grus grus* observed by radar and from an aircraft. *Ornis Scandinavica*, **10**, 241–51.

Pennycuick, C. J., Obrecht, H. H., and Fuller, M. R. (1988). Empirical estimates of body drag of large waterfowl and raptors. *Journal of Experimental Biology*, **135**, 253–64.

Prandtl, L. and Tietjens, O. G. (1934*a*) *Fundamentals of hydro- and aero-mechanics*. United Engineering Trustees. (Dover edition 1957.)

Prandtl, L. and Tietjens, O. G. (1934*b*). *Applied hydro- and aero-mechanics*. United Engineering Trustees. (Dover edition 1957.)

Prior, N. C. (1984). *Flight energetics and migration performance of swans*. Ph.D. Thesis, University of Bristol.

Rayner, J. M. V. (1979). A vortex theory of animal flight. Part 1. The vortex wake of a hovering animal. *Journal of Fluid Mechanics*, **91**, 697–730.

Rayner, J. M. V. (1985). Bounding and undulating flight in birds. *Journal of Theoretical Biology*, **117**, 47–77.

Rayner, J. M. V. (1986). Vertebrate flapping flight mechanics and aerodynamics, and the evolution of flight in bats. *Biona Report*, **5**, 27–74.

Rothe, H. J., Biesel, W., and Nachtigall, W. (1987). Pigeon flight in a wind tunnel. II. Gas exchange and power requirements. *Journal of Comparative Physiology*, **157**, 99–109.

Rothe, H. J. and Nachtigall, W. (1987). Pigeon flight in a wind tunnel. I. Aspects of wind tunnel design, training methods and flight behaviour of different pigeon races. *Journal of Comparative Physiology*, *B*, **157**, 91–8.

Schaffner, F. C. (1988). *The breeding biology and energetics of the white-tailed tropicbird (Phaethon lepturus) at Culebra, Puerto Rico.* Ph.D. Thesis, University of Miami.

Schmitz, F. W. (1960). *Aerodynamik des Flugmodells* (4th edition). Lange, Duisberg.

Spedding, G. R. (1981). *The vortex wake of flying birds: an experimental investigation.* Ph.D. Thesis, University of Bristol.

Spedding, G. R. (1986). The wake of a jackdaw (*Corvus monedula*) in slow flight. *Journal of Experimental Biology*, **125**, 287–307.

Spedding, G. R. (1987). The wake of a kestrel (*Falco tinnunculus*) in flapping flight. *Journal of Experimental Biology*, **127**, 59–78.

Spedding, G. R., Rayner, J. M. V., and Pennycuick, C. J. (1984). Momentum and energy in the wake of a pigeon (*Columba livia*) in slow flight. *Journal of Experimental Biology*, **111**, 81–102.

Tucker, V. A. (1973). Bird metabolism during flight: evaluation of a theory. *Journal of Experimental Biology*, **58**, 689–709.

Tucker, V. A. (1987). Gliding birds: the effect of variable wing span. *Journal of Experimental Biology*, **133**, 33–58.

Tucker, V. A. (1988) Gliding birds: descending flight of the white-backed vulture, *Gyps africanus*. *Journal of Experimental Biology*, **140**, 325–44.

Tucker, V. A. and Parrott, G. C. (1970). Aerodynamics of gliding flight in a falcon and other birds. *Journal of Experimental Biology*, **52**, 345–67.

Vogel, S. (1981). *Life in moving fluids: the physical biology of flow*. Princeton University Press, Princeton.

White, D. C. S. and Thorson, J. (1975). *The kinetics of muscle contraction*. Pergamon, Oxford.

Williams, T. C., Williams, J. M., Ireland, L. C., and Teal, J. M. (1977). Autumnal bird migration over the western North Atlantic Ocean. *American Birds*, **31**, 251–67.

Index and glossary of scientific names